331.4815 C61 2221822
CLIMBING THE LADDER

Climbing the Ladder

An Update on the Status of Doctoral Women Scientists and Engineers

Committee on the Education and Employment of Women in Science and Engineering

Office of Scientific and Engineering Personnel

National Research Council

NATIONAL ACADEMY PRESS
Washington, D.C. 1983

Library of Congress Catalog Card Number 83-60184

International Standard Book Number 0-309-03341-1

Available from:

NATIONAL ACADEMY PRESS
National Academy of Sciences
2101 Constitution Avenue, NW
Washington, DC 20418

Printed in the United States of America

COMMITTEE ON THE EDUCATION AND EMPLOYMENT OF WOMEN IN SCIENCE AND ENGINEERING

Lilli S. HORNIG, Chair
Executive Director, Higher Education Resource Services
Wellesley College

M. Elizabeth TIDBALL, Vice-Chair
Professor of Physiology
George Washington University Medical Center

John A. ARMSTRONG
Manager of Materials and Technology Department
IBM Corporation

Esther M. CONWELL
Principal Scientist
Xerox Corporation

Eleanor I. FRANKLIN
Professor of Physiology
Howard University College of Medicine

Gertrude S. GOLDHABER
Senior Physicist
Brookhaven National Laboratories

Dudley R. HERSCHBACH
Professor of Chemistry
Harvard University

Shirley A. JACKSON
Theoretical Physicist
Bell Laboratories

Vera KISTIAKOWSKY
Professor of Physics
Massachusetts Institute of Technology

Barbara F. RESKIN
Associate Professor of Sociology
Indiana University

David Z. ROBINSON
Vice President
Carnegie Corporation of New York

Elizabeth L. SCOTT
Professor of Statistics
University of California, Berkeley

M. Lucius WALKER, Jr.
Dean, School of Engineering
Howard University

Nancy C. AHERN, Staff Officer

PREFACE

This is the fourth report of the Committee on the Education and Employment of Women in Science and Engineering and its second publication to address specifically the status of women scientists and engineers in academic institutions. The present report updates an earlier study, _Climbing the Academic Ladder: Doctoral Women Scientists in Academe_, and examines any changes in the status of women faculty between 1977 and 1981. Drawing on more limited data, it also considers their situation in industry.

The Committee believed that another look at the situation, after four years, was needed for several reasons. The supply of doctoral women scientists and engineers grew sharply in the late 1970s. At the same time that the numbers of women leaving graduate school had increased significantly, the numbers of positions in colleges and universities had not. Whether women are affected disproportionately by the relative scarcity of academic jobs needed to be examined. It was also of interest to determine whether the gains in the presence of women on science and engineering faculties noted during the mid-1970s had been sustained. The Committee wished to examine the extent to which salary differentials observed in earlier studies had narrowed or perhaps disappeared for men and women with recent doctorates. Finally, the status of doctoral women scientists and engineers in industry needed to be reviewed.

Data for the report were obtained primarily from the Survey of Doctorate Recipients and Survey of Earned Doctorates conducted by the National Research Council under contract with four federal agencies.

Financial support for this study was provided by the Ford Foundation and is most gratefully acknowledged. Mariam Chamberlain and Gladys Chang Hardy, who served successively as the Foundation's staff officers, were especially helpful. The IBM Corporation generously awarded a supplemental grant for an analysis of women scientists in industry which appears in Chapter 5 of this report. A. N. Scallon's assistance at IBM is noted with thanks.

Since its inception the Committee has been chaired by Lilli S. Hornig, Executive Director, Higher Education Resource Services, Wellesley College. For this and each of the Committee's previous reports, Nancy C. Ahern served as staff officer. Judith F. Vassalotti as secretary to the Committee typed the text of this report and prepared the numerous tables and figures. William C. Kelly, formerly Executive Director of the CHR and now Executive Director of the Office of Scientific and Engineering Personnel, has contributed advice, editorial comments, and much time and wisdom to each of the Committee reports. We take this opportunity to express particular thanks to him.

CONTENTS

LIST OF TABLES

LIST OF FIGURES

INTRODUCTION

The Committee's first report, Climbing the Academic Ladder: Doctoral Women Scientists in Academe (1979), explored the status of women in faculty, postdoctoral, and advisory posts. Women scientists were found to be concentrated in the lower ranks and in off-ladder positions, were typically paid less than their male colleagues at the same rank, and were less likely to be awarded tenure. A subsequent report, Career Outcomes in a Matched Sample of Men and Women Ph.D.s: An Analytical Report (1981), indicated that these differences remain even when men and women are closely matched by education, experience, and type of employment. The latter study also revealed that the disparities in pay and advancement are not explained by what are traditionally considered important factors--the perceived greater restraints on career mobility or greater likelihood that women have in the past interrupted their careers for child-rearing.

Since 1977--the survey year on which the first report was based--an additional 13,000 doctoral women scientists and engineers have joined the labor force, bringing their number to 41,000 of a total of 341,000 for both sexes. The characteristics of the new entrants and their effect on the overall status of women in academe and industry are examined in the following analyses.

Data sources

The primary data on which this report is based are a pair of surveys conducted by the National Research Council. Copies of the questionnaires are provided in Appendices A and B.

The annual Survey of Earned Doctorates is a virtually 100 percent survey of individuals receiving doctorates from U.S. institutions. With the assistance of graduate deans, information is collected at the time of receipt of the Ph.D. concerning educational background and postdoctoral plans. These data are analyzed in Chapter 2, "Supply of Women Doctorates."

The follow-on Survey of Doctorate Recipients collects subsequent employment data from a sample of 65,391 scientists, engineers, and humanists who earned Ph.D.s during the period 1938-1980. The survey has been conducted biennially since 1973; this report relies chiefly on the 1981 survey results. Responses from individuals in the sample are weighted to yield population estimates. The estimates are in turn subject to possible error due to sampling variability and possible nonsampling errors such as nonresponse bias.

Organization of the report

We will first look at the number of women in the science/engineering pipeline and recent trends in the proportion of college women planning careers as scientists. In Chapter 2, the characteristics of new doctorate recipients are described. Chapter 3 presents data on patterns of postdoctoral appointments for recent Ph.D.s. The comparative status of men and women faculty, including their rank, tenure, and salary profiles, is discussed in Chapter 4. Chapter 5 briefly examines the employment patterns of doctoral women in industry and whether the picture has changed since 1977. Finally, we summarize the evidence related to the status of women scientists and propose recommendations for improving their situation.

SUMMARY OF FINDINGS

- Undergraduate science enrollments for women have increased steeply, especially in those fields in which women have been most under-represented--engineering and computer sciences. (Table 1.1)

- The persistence rate or probability of women going on to graduate school is low, relative to men, in mathematics and chemistry. (Table 1.2)

- The numbers of women earning Ph.D.s in science and engineering have increased steadily since 1970 while the numbers of men have declined. The decade of the 1970s was the first in which the percent of doctorates granted to women matched or exceeded the levels of the 1920s. (Figure 2.1 and Table 2.1)

- In the 5-year period 1976-1980 similar proportions of men and women doctorates had received their training at highly rated departments. The sex differences were small except in mathematics where women were less likely to have received their degrees from prestigious institutions. (Table 2.2)

- Except in computer sciences and physics, 15-20 percent more women than men were still seeking jobs. In general, a somewhat higher proportion of the male Ph.D. recipients reported having definite jobs at the time of graduation except in computer sciences. (Table 2.4)

- Women constituted 12 percent of all doctoral scientists and engineers in the labor force in 1981. (Table 2.6)

- Similar proportions of recent men and women Ph.D.s planned post-doctoral study. (Table 3.1)

- About one-fifth of all postdoctorals have held long-term appointments--for more than 36 months. The holding pattern is most prevalent in the life sciences, especially for married women and single men. (Table 3.4)

- Married women report that geographic limitations played an important role in their decision to take a postdoctoral appointment. (Table 3.5)

- As of 1981, there were approximately 13,500 doctoral women on U.S. science and engineering faculties, accounting for 10.9 percent of the total. Their representation is up from 9.3 percent in 1977. (Table 4.2)

- In the major research universities, women held 24 percent of the assistant professorships, but only 3 percent of the full professorships as of 1981. (Table 4.3)

- Women scientists are still twice or three times as likely as men to hold nonfaculty (instructor/lecturer) appointments. In most fields, the disparity has increased since 1977. However, relatively few (3 percent) of all doctoral women in academe hold such positions. (Table 4.4)

- In general, recent women Ph.D.s are found in junior faculty positions in proportions exceeding their availability in the doctoral pool. (Table 4.6)

- Promotions of junior faculty between 1977 and 1981 show wide sex differences: in the group of top 50 institutions (ranked by R&D expenditures), for example, three-fourths of the men, but only one-half of the women were promoted from assistant professor to a higher rank in those years. (Figure 4.3)

- Overall, the proportion of women scientists who are tenured continues to be lower than for men. The sex differential in tenure for associate professors, however, has narrowed since 1977, and, at the assistant professor rank, a slightly higher percent of the women have tenure. (Table 4.7)

- The median "time to tenure," for science and engineering faculty who did achieve tenure, was 5.9 years for women and 6.1 years for men. In the physical sciences, however, awarding of tenure lagged for female faculty. (Table 4.8)

- After controlling for rank, salary differences for men and women persist in most fields. The sex differences continue to be largest in chemistry and the medical sciences. (Table 4.10)

- The number of doctoral women scientists and engineers in industry doubled between 1977 and 1981. Still, women account for only 5 percent of all Ph.D.-level industrial personnel. (Table 5.1)

- New women Ph.D.s now plan industrial employment at about the same rate as men. (Table 5.2)

- Women scientists and engineers continue to be less likely to hold managerial jobs in industry. (Table 5.3)

- Median salaries in industry are typically lower for female scientists, even among the recent Ph.D.s. For those 1-2 years past the doctorate, the salary gap amounts to $2,400. (Figure 5.2)

CHAPTER 1

FACTORS AFFECTING THE SUPPLY OF DOCTORAL WOMEN SCIENTISTS AND ENGINEERS

Women have traditionally constituted a small minority of all doctoral scientists and engineers, although their representation varies considerably by field across the spectrum of scientific and technological disciplines. During the decade of the 1970s, for example, women earned about a quarter of all social science doctorates and one-fifth of those in life sciences but less than one-tenth in physical sciences and only 2 percent in engineering. The proportion of women in all these fields has been rising, dramatically in some areas, but despite this trend women's relatively low apparent interest in science fields remains a central issue in any assessment of women's status in the sciences. Although the dominant concern of this report is to examine the progress in academic careers of those women who do choose to pursue science, it is necessary also to explore the factors that may affect such a choice for good or ill throughout the course of preparation. In this chapter we will examine both individual and structural factors that contribute to educational and career choices.

Historical patterns and their reflections

In a general framework of human resources studies there are two basic preconditions for achievement in science or any other calling: ability and opportunity. The interactions between these two factors can lead to many different results, but two boundary conditions exist in the real world: 1) even outstanding ability needs to be trained, directed, provided with a sphere of action, and rewarded in order to come to fruition--in short, to have opportunity to flourish; and 2) even unlimited opportunity cannot create achievement where requisite ability does not exist. Historically, this quite obvious generalization has been seen as valid for men, leading especially in the sciences in the period following World War II to the creation of an elaborate system of educational and career support opportunities (pre- and postdoctoral fellowships, career service awards, research support, etc.) predicated on two assumptions: 1) that whatever scientific talent exists initially must be nurtured by enhancing appropriate opportunities, and 2) that it is in the national interest to do so.

In the case of women's careers in the sciences, however, these simple assumptions have not been applied in the same way, with the result that both women's access to training in science and the later development of their careers have been forced into a different mold. From the mid-19th century when women began to be admitted to higher education up to the late 1960s, the dominant rationale for educating women at all was that it would make them better wives for educated men, and better mothers for their children. True professional career preparation, especially in the sciences, became a reality only for a handful of outstanding women in each generation. Some of the factors that still affect adversely women's access to and development in science careers are directly traceable to these limited perceptions.

Aside from the issues of potential sex differences in cognitive ability discussed more fully below, women's aspirations and achievement motivation have been examined and questioned from a perspective that assumed the absolute primacy of women's "instinctive" desire to marry and raise children without evaluating how realistic their career-related motives and aspirations might indeed be in light of the actual opportunities open to them. In contrast to this position, the Committee makes only one assumption in examining the factors that affect women's achievement in science: that all individuals, male or female, make educational and career choices in a way that they believe will maximize their chances for a productive, rewarding, and happy life within the framework of opportunities they perceive as real. Such an assumption has the advantage of enabling us to interpret rationally the recent trends of steeply increasing participation of women in science as a normal and expected response to expanding opportunities, rather than having to postulate a dramatic, fundamental shift in female psychology occurring in the 1970s. In the remainder of this chapter we examine evidence relating to cognitive ability, opportunities for access to science careers, and how the interactions between them may affect men and women differently.

We use a "pipeline/valve" model of the education-career sequence as a conceptual framework for this examination. In such a generalized model, the valves represent successive selection processes in attainment, such as represent successive selection processes in attainment, such as undergraduate and graduate degrees, junior faculty appointments, and promotion to tenure. Among other properties, such a system has the characteristic that if the final valve in the line is constricted even if all others remain open, the flow will decrease. Flow will be dependent on both external factors such as availability of places in graduate school, financial support for students, and professorial appointments, and on internal or self-selection of individuals who assess by whatever means and information accessible to them how good their chances are of reaching a desired stage. Women's decisions to pursue or abandon careers in science will thus be evaluated against the background of real or perceived opportunities open to them at several periods in time. Note that perceived opportunities are equally if not more important than real ones, since individuals make choices based on their perception of a situation, which may differ from the facts.

Scientific ability

The assumption that women's general intellectual endowment is inferior to men's is an ancient one which was elaborated during the 19th century by such diverse disciplines as psychology (Shields, 1975) and physical anthropology (Gould, 1981) and publicly proclaimed on innumerable occasions during the debates surrounding women's admission to higher education (Conable, 1977; Earnest, 1953). More recently, this assumption has become less sweeping, restricted primarily to putative sex differences in mathematical ability; if women's general mathematical ability or some relevant segment of it can be shown to be inferior to men's, such a finding would explain low participation in science, especially in the most quantitative areas. It has been suggested that differences in mathematical ability may correlate with differences in brain lateralization (McGee, 1979), and most recently in androgen levels (Hier and Crowley, 1982). In the latter case, low spatial ability and low androgen levels were shown to coincide in males, although no similar quantitative correlations were attempted for females. It is perhaps too early to tell whether such conjectures are as erroneous as the 19th century focus on sex differences in brain size, which is now recognized to be related to body size (Gould, 1981) rather than intelligence. Kagan (1982) has pointed out a number of fallacious assumptions in Hier and Crowley's work.

The information in the literature that bears on sex differences in scientific ability is restricted largely to aspects of mathematical ability, which are used as a proxy measure. The connections between such discrete factors as spatial visualization, abstract reasoning, or analytical ability (which have been tested) and successful scientific achievement seem intuitively obvious to most scientists, but the nature of the relationships have not been analyzed and tested.

In broad outline, well-defined sex differences are found in scores on large-scale standardized tests such as the Scholastic Aptitude Test - Math (SAT-M) and the National Assessment of Educational Progress (NAEP), with girls on the average scoring lower than boys on the SAT-M by about 0.4 standard deviations. Most but not all of the difference disappears when the results are controlled for the number of math courses taken in high school and for whether or not individuals are enrolled in a math course at the time of the test; the test therefore almost certainly represents a measure of achievement rather than innate aptitude. When test results are properly controlled for course-taking, a small difference favoring males remains in segments testing problem-solving for those students who have not taken calculus or geometry but no difference is found between boys and girls who have taken these courses (Armstrong, 1979).

A number of studies (reviewed by Lockheed, 1982) have examined the possibilities of sex bias inherent in a variety of standardized tests, primarily in terms of either content bias or psychometric bias. Such types of bias have not been demonstrated conclusively, and Lockheed concludes that the differences in scores represent real sex differentiation in learning experience.

Sex differences in either mathematical performance or interest are not observed in young children but differences in expressed interest appear among academically highly able students in 7th or 8th grade, with boys much more interested in math. This greater interest appears to be closely linked to career interests in scientific fields (Haven, 1972). Other findings (Astin, 1968; Fox, 1981; Helson, 1971) support the conclusion that timely encouragement during adolescence is important in sustaining mathematics-related interests of girls and boys, but that girls are far less likely to receive it from any source.

In contrast to findings of lower mean scores for women on national tests, other researchers (Fennema and Sherman, 1977) find that in certain schools or classes no sex differences in mathematics achievement can be demonstrated concluding that this outcome results from absence of various kinds of sex bias in such schools.

In any event it is clear that from the point at which math and science courses become elective, girls have in the past chosen (or were allowed) to participate less than boys, resulting in female high school graduates having had about one-half course less than males in these fields (National Center of Education Statistics, 1979); to what extent the courses girls do take may differ from those taken by boys (e.g., business math versus algebra II) is not clear. In the aggregate, sex differences in high school science and math preparation or in existing aptitude measures appear to be too small to account readily for the disparate distributions of women and men in these fields in college. For example, 46.4 percent of male high school students versus 43.1 percent of females take college preparatory math (including algebra II or III, geometry, trigonometry, pre-calculus, and calculus) and 21.8 percent of boys versus 17.9 percent of girls take chemistry or physics (NCES, 1981). Precise information on the high school sex distributions in calculus or advanced chemistry and physics courses is not available.

One qualitative study of high school girls in Advanced Placement math and science courses (Casserly, 1979) reports pervasive efforts by counselors to steer girls out of these courses, chiefly on grounds that they would have to work too hard or might "spoil a good record." Male counselors who tried to dissuade girls from taking such courses are quoted as worrying that these girls might take needed jobs away from men. One director of guidance stated (p. 12): "There are men with Ph.D.s in physics all over the place who can't get jobs. Why should we encourage girls? Why, if they're successful, they'd be taking jobs away from men who need them. No, it wouldn't be fair to the girls."

Nonetheless, several studies in the late seventies suggest that sex differences in course taking are disappearing (cited in Fox, 1980, p. 8) as a result of increased participation by girls. Fox concludes that these changes are still in progress but are occurring faster in some age, ability, and/or geographic groups which may account for variation in findings. Such changes should be reflected in increased participation at the college level within the next two or three years and by the late eighties at the Ph.D. level.

Trends in math and science preparation in college

Student distributions by sex among various college majors are customarily measured as percent of baccalaureate degrees earned, a misleading indicator which does not take account of the changing sex composition of the total student body over time. For example, between 1970 and 1980 the proportion of total baccalaureates earned by women rose from 43.2 percent to 49 percent, with two-thirds of the increase occurring in the second half of the decade. The growth in women's participation in science is measured more usefully as the ratio of the percent degrees earned in a given science field to the percent of total baccalaureates earned by women in a given year, as in Table 1.1. This ratio also has the characteristic that it approaches unity as the sex distributions converge, and thus constitutes an accurate measure, termed "parity index," of the relative interests of college men and women in a given field at a particular time.[1]

Parity indices for female participation in selected broad fields of college science are given in Table 1.1. Steeply increasing indices are noted in those fields in which women have been most underrepresented, i.e., engineering and computer sciences. Mathematics, long the most sex-neutral of college majors, shows a very slight total decline in popularity among women over the last decade after peaking in the early 70s, while during the same period women's relative participation in physical sciences has increased by about 55 percent. In biological sciences, historically the most popular of the natural sciences among women, participation over the decade has risen by about one-third and is now approaching the male distribution.

[1]The prior question posed by using this measure--the determinants for both sexes of choosing to attend college at all, and choosing which college to attend--is beyond the scope of the present report but is addressed fully in Perun, 1982. Significant sex differences in financial support generated by the G.I. Bill in the post-World War II era account to a large extent for sex differences in college attendance.

TABLE 1.1 Growth in baccalaureate degrees to women in science and engineering fields,[a] 1960-1980

Year	Women as % of all BAs	Math		Physical sci.		Biological sci.		Engineering		Computer sci.	
		% BA[b]	PI[c]	% BA	PI	% BA	PI	% BA	PI	% BA	PI
1960	35.0	27.2	.78	12.5	.31	25.2	.72				
1962	37.4	29.1	.78	13.4	.36	28.3	.76				
1964	40.0	32.0	.80	13.8	.35	28.2	.71				
1966	40.3	33.3	.83	13.5	.33	28.0	.69				
1968	41.4	37.1	.90	13.6	.33	27.8	.67				
1970	43.2	37.4	.87	13.6	.31	27.8	.64	0.7	.02		
1972	43.7	39.0	.89	14.9	.34	29.4	.67				
1974	44.4	40.9	.92	16.5	.37	31.2	.70				
1976	45.5	40.7	.89	19.2	.42	34.6	.76	3.2	.07	13.6	.31
1978	47.1	41.3	.88	21.5	.46	38.7	.82	6.6	.14	25.7	.55
1980	49.0	42.3	.86	23.7	.48	42.1	.86	10.1	.21	30.2	.62

[a] Comparable data for social sciences are not available.

[b] %BA = women as percent of all BAs in given field.

[c] $PI = \frac{\text{\% women BAs in field}}{\text{\% of all BAs to women}}$

SOURCE: Compiled from National Center for Education Statistics data.

Table 1.1A, showing trends in women's participation in English, foreign languages, and education majors is included for comparison with the trends in science fields. The proportional increases in the popularity of science fields among women are comparable to the decreases in these non-science fields; however, it should also be noted that the numbers involved are far larger in education (nearly seven times as large as physical sciences) and English (about twice as large). The numbers of foreign language majors are about the same as mathematics. These opposite trends in field distributions for women obviously do not reflect a simple trade-off, but rather some very broad shifts in both expanding options and interests of women students.

These measures demonstrate a real growth of interest in science fields among women beyond the increases deriving simply from the fact that the proportion of total baccalaureates earned by women has grown by about 13 percent over the decade. In terms of the pipeline model, some previously constricted valves have been opened. The causes of this real increase have not yet been addressed in focused research studies. One possible causal factor would be better high school preparation of women students stimulated by a variety of career information and workshop programs during the 70s, many of them supported by the National Science Foundation but now eliminated on budgetary grounds. Heightened aspirations and career expectations fostered by the perception of sex equality generated in the last fifteen years can be assumed also to be an important contributing factor; the expectation that women's career rewards in the sciences will approximate or equal men's did not realistically exist much before 1970. Certainly the near equalization of financial support for both sexes that has occurred as a consequence of equal rights statutes (specifically Title IX of the 1972 Higher Education Act and the Equal Credit Act of 1974) has made a significant contribution to equalizing college attendance and possibly has prompted women to choose more often those fields in which graduate work is also likely to be necessary for full professional development.

Trends in graduate education

The decision to undertake graduate training represents an important valve in our pipeline model; if men and women base these decisions on different information or if the same information has different significance for the two sexes, then the decisions can also be expected to differ. In the preceding section we have examined how women's initial undergraduate field choices have changed in the last two decades to approach the male patterns of field distribution more closely. In this section we will examine to what extent women's decisions to pursue Ph.D.s in the sciences exhibit a parallel trend. To assess how much variation needs to be explained, it is instructive to look at both persistence/attrition rates and at a measure somewhat analogous to the parity indices described in the previous section.

Table 1.1A

Trends in Proportions of Baccalaureate Degrees

Earned by Women in Selected Humanities Fields and Education, 1960-1980

Year	Women as Percent of all BAs	English Percent BA	English PI[a]	Foreign Lang. Percent BA	Foreign Lang. PI	Education Percent BA	Education PI
1960	35.0	62.3	1.78	65.8	1.88	71.1	2.03
1962	37.4	64.9	1.74	68.6	1.83	73.2	1.96
1964	40.0	66.4	1.66	72.7	1.82	76.2	1.91
1966	40.3	66.2	1.64	73.0	1.81	75.4	1.87
1968	41.4	67.3	1.63	74.6	1.80	75.9	1.83
1970	43.2	66.9	1.55	74.7	1.73	75.0	1.74
1972	43.7	65.8	1.51	75.5	1.73	74.1	1.70
1974	44.4	63.9	1.44	76.6	1.73	73.5	1.66
1976	45.5	62.6	1.38	76.8	1.69	72.8	1.60
1978	47.1	63.6	1.35	76.4	1.62	72.5	1.54
1980	49.0	--		75.9	1.55	73.2	1.49

a. $PI = \frac{\text{\% Women BAs in field}}{\text{\% of all BAs to Women}}$

Source: Compiled from National Center for Education Statistics data.

TABLE 1.2 Persistence and attrition of women from the science and engineering educational ladder

	% Women among:			
	Entering college freshmen, by probable major[a] 1968-69	Bachelor's degrees awarded[b] 1971-72	Master's degrees awarded[b] 1973-74	Doctoral degrees awarded[c] 1978-79
Mathematics/statistics	45.9	39.1	31.0	15.4
Computer sciences	---	13.6	12.9	12.9
Physical sciences	14.7	15.1	14.6	11.5
Physics	---	6.9	8.2	6.6
Chemistry	---	19.4	21.8	14.0
Engineering	1.3	1.0	2.3	2.5
Agriculture	2.0	5.5	9.8	9.0
Biological sciences	36.7	29.6	30.6	26.8
Social sciences	63.3	38.5	31.8	33.0

[a] Derived from statistics published in National Norms for Entering College Freshmen, Fall 1968, American Council on Education.

[b] As reported in the series Earned Degrees Conferred, National Center for Education Statistics.

[c] Summary Report 1979: Doctorate Recipients from U.S. Universities, National Research Council.

1. Persistence and attrition

The perception that women are far less likely than men either to begin graduate work at all or to complete a doctorate in science fields is very widespread, and was widely used in the past to justify denial of financial support to women graduate students (Bernard, 1964), thereby generating a self-fulfilling prophecy. One useful way of measuring suspected attrition is shown in Table 1.2, which compares the percent women among probable majors of entering college freshmen in 1968-69 with baccalaureate, master's, and doctoral degrees earned at appropriate subsequent intervals. Some field differences are apparent at once in the persistence rate through college: prospective majors in social sciences suffer a large decline by graduation, and a modest decline occurs in both mathematics and biological sciences, while the physical sciences and agriculture actually show a relative gain. At the master's degree level, significant attrition has taken place in mathematics and social sciences but several fields show slight gains, and in agriculture the likelihood of women's earning the degree is increased dramatically. With respect to the doctorate, by far the largest decrease occurs in mathematics, for an attrition rate of over 60 percent from the BA, most of which occurs after the master's degree. In chemistry the attrition rate is also large, amounting to about 25 percent from the baccalaureate. Biological and social sciences both show relatively small decreases between the baccalaureate and doctorate, while the probability of women's earning a Ph.D. stays essentially constant compared to men in computer sciences and physics but rises in engineering and agriculture.

Such marked differences by field in persistence to the Ph.D. between men and women are not readily explained in any simple way. They do not correlate well, for example, with various measures used by Feldman (1974) to diagnose perceptions of sex equity for graduate students in various disciplines. This study provides evidence that in a large-scale national survey carried out in 1969, about 15-35 percent of both male and female graduate students in science fields believed that faculty do not take women seriously, a belief generally corroborated by the corresponding faculty survey. However, this opinion was much less pronounced for mathematics than for chemistry, physics, most biological sciences, and several social science fields.

The observed variations in persistence rates, both among fields and between the sexes, suggest that the nature of the choices made by men and women continues to be determined by somewhat different factors. In our pipeline model, there are several possible branches at each valve or degree level, such as immediate employment, a switch to a different academic field, a switch to professional training such as law or medicine, and others. "Attrition" between the baccalaureate and the doctorate in certain fields (biosciences, some social sciences, chemistry) may, for example, partially reflect the very large increases in medical school attendance by women that followed the lifting of quota restrictions in these schools in the early 70s.

No clear explanation exists for the marked discrepancy in persistence rates between mathematics and other science fields; there is a resemblance to chemistry in that most of the drop occurs after the master's degree. Mathematics, however, is a small field compared to most other sciences. It is perhaps more important to focus attention on the high persistence rates in other science fields.

2. "Parity" at the doctoral level

Since a detailed discussion of women doctorates in science and engineering is presented in Chapter 2, this section will confine itself to assessing the degree of parity women have achieved at the Ph.D. level. It should be emphasized that the percentages of degrees earned by women in each field are the important determinant of their representation in the labor pool, and therefore are the appropriate measure for career progress. Hence the actual proportions of degrees earned by women will serve as the reference standard throughout the remaining chapters of this report. In this section, however, we apply two different measures to assess separately 1) the extent to which women's field distribution in a given Ph.D. cohort approximates that of men, and 2) the propensity of women baccalaureates in a given BA cohort relative to corresponding men to earn a Ph.D. after a time interval appropriate to the particular field. These measures, respectively termed PI_1 (parity index) and PI_2, are presented in Table 1.3, comparing results for 1970 and 1980 Ph.D.s.

Table 1.3 shows that in 1970, when we consider the field distributions of all doctorates, women were slightly more likely than men to have earned the degree in biosciences and much more so in psychology, but much less so in mathematics and least of all in physical sciences (PI_1). With reference to the relevant baccalaureate pool at a field-specific appropriate earlier time (PI_2), however, a somewhat different picture emerges, comparable to the persistence rates discussed above. Women baccalaureates in mathematics were only about one-quarter as likely as their male colleagues to proceed to a doctorate and in psychology that probability was 60 percent, with the other fields falling between these extremes. The reader should note that these ratios for mathematics are not highly reliable because they are based on numbers smaller than 100 (see Note c, Table 1.3).

By 1980 no change is observed in PI_1 for physical sciences and the mathematics value has decreased, apparent evidence of a diverging sex distribution pattern in that field. The values for both biosciences and psychology are reduced compared to 1970. A dramatic change has occurred, however, in PI_2 in all fields except mathematics; based on the relevant BA pools, there is a close approach to parity in biosciences and psychology, and the parity measure has nearly doubled in physical sciences to the .75 level, but has increased only slightly in mathematics.

TABLE 1.3 Parity trends for Ph.D.s by broad field, 1970 and 1980

	1970			1980		
	% Ph.D.s in field to women	PI_1[a]	PI_2[b]	% Ph.D.s in field to women	PI_1	PI_2
Mathematics	(7.8)[c]	(.59)	(.27)	(12.8)	(.42)	(.33)
Physical sci.	5.4	.40	.39	12.3	.41	.75
Biological sci.	14.3	1.08	.51	28.0	.92	.94
Psychology[d]	22.3	1.68	.60	42.3	1.40	.97

[a] $PI_1 = \frac{\text{Percent women Ph.D.s in field}}{\text{Percent women Ph.D.s in all fields}}$

[b] $PI_2 = \frac{\text{Percent women Ph.D.s in field}}{\text{Percent women BAs in field y years earlier}}$

where y = 6 years for physical sciences
= 7 years for biological sciences
= 8 years for mathematics and psychology

[c] Bracketed figures indicate percentages based on absolute numbers of Ph.D.s below 100, where indices become very unstable due to fluctuations.

[d] Because of early differences in data collection, aggregate figures for the broad field of social sciences are not available.

SOURCE: Computed from National Center for Education Statistics, 1979 and 1980; National Academy of Sciences, 1981.

These patterns are clear evidence of strong convergence of men's and women's educational choices and decisions in all science fields except mathematics. It should be noted that a similarly converging pattern is developing in engineering and computer sciences, although the numbers of women Ph.D.s are too small to warrant separate presentation here. These distribution ratios demonstrate among other things the importance of taking account of general education history when we attempt to explain women's low participation in science; in particular, the size of the appropriate baccalaureate pools has a marked influence on determining the degree of parity to be expected. It seems evident that until women achieve general parity at the baccalaureate level (i.e., 52 percent of all BAs), an event expected within the next five years or so, representation parity at the Ph.D. level cannot occur. Parity at the BA level (which was almost achieved just before World War II, although the data have not been presented here) depends primarily on two factors: the absence of formal sex distinctions in admissions and equality of access to financial support. Unless these conditions can be met, it is futile to expect full participation of women in science, and hence to have available an expanded pool of scientists to fill national needs. While formal admissions quotas for women have been eliminated in most institutions, the situation with respect to equal financial support may be more problematic. In the years immediately after World War II, the proportion of male college students rose from 55 percent to over 70 percent, and about 70 percent of all the men were supported by the G.I. Bill (Olson, 1974). As a result of this subsidy, higher education became accessible to men with the requisite ability regardless of their economic status, while women had to continue to finance their own education, thus in effect restricting the numbers who could attend college. This situation was exacerbated by the common practice of awarding women less financial aid from institutional sources as well; as late as 1969-70, institutionally-administered grants to male college students averaged $671 and those to women only $515 (Haven and Horch, 1972), a difference of 32 percent. Extensions of the G.I. Bill for Korean and Vietnam war veterans had effects similar to the original bill but much reduced in total numbers and value.

If the efforts to increase military recruiting through the promise of future educational benefits that are now under way in the Congress prove successful, such a policy will again adversely affect women's relative access to higher education and hence to science careers. Even if future enlistments in the military forces remain open to women, it is unlikely that the present low quotas and higher qualifications required of women will be changed: on the other hand, there is reason to believe that scientifically talented young people of either sex would not enlist in significant numbers except on the case of a national emergency. Nonetheless, educational benefits as a reward for military service would again produce an artificial long-term dominance of men in higher education, such as occurred in the period following World War II, and hence revived the damaging perception that women are somehow unfit for advanced work.

Therefore this committee recommends that the Congress explore measures to ameliorate the inevitable sex bias in higher education that will again result from a renewed G.I. Bill. The purpose of raising this question is to emphasize our concern not simply with the numerical disparities that result from a sex-biased support pattern but particularly with the consequent limitations on access for talented women from less than affluent families.

The increases in women's propensity to prepare for careers in science and engineering over the last decade are too rapid to have been caused by some fundamental change in either female aptitudes or psychological makeup. Such a change is likely to require a far longer time period. It is much more probable that the increases have occurred as the result of two related factors: the ending of formal sex discrimination in institutions of higher education, in large part by the various civil rights statutes, and the consequent perception by women that since these statutes guarantee a measure of equality it is now relatively far more rewarding to make the personal and financial investment that is required to prepare for careers in science and technology.

The PI_2 ratios allow us to make some modest predictions of the proportions of women to be expected in Ph.D. pools in the future. If we assume that the rates at which women perceive science fields to offer an equitably rewarding future do not change from the 1980 levels, we would expect about 18 percent women doctorates in physical sciences in 1986 and about 40 percent in biosciences in 1987, but only 14 percent in mathematics by 1988. If women's propensity to complete doctorates in physical sciences continues to rise, on the other hand, at the same rate as in the recent past, women in these fields would attain statistical parity in the late 1980s at about 25 percent of Ph.D.s. Numerical parity is likely to be attained only after it has occurred at the baccalaureate level, which would happen in about the mid-1990s at the current rate of growth. The fact that science participation by high school girls is currently increasing gives reason to believe, however, that the growth of the female BA pool will accelerate in the near future, other factors remaining equal.

The situation in two fields of great current interest, engineering and computer sciences, is still much more uncertain. In engineering women started from a near-zero base a decade ago and were 10 percent of baccalaureates and 3.6 percent of Ph.D.s in 1980. The high growth rates for women undergraduate engineers became most marked in the last few years and will not be reflected among Ph.D.s until the late-1980s. In computer sciences (where separate records are not available before 1971) women BAs were 13.6 percent in 1971 and 30.2 percent in 1980, but the proportion of Ph.D.s has remained essentially constant at about 9.5 percent. It is probable that in this field also the rate of growth will accelerate during the 1980s, based on baccalaureate figures.

Two kinds of conclusions emerge from these considerations. The first is that although the participation rates of women in mathematics graduate degrees remain something of a puzzle, in most science fields the flow through the pipeline is accelerating at a rate that leaves little doubt of either the fitness or inclination of women to undertake professional careers in these disciplines. We have noted elsewhere (see Chapter 2 and also Climbing the Academic Ladder, pp. 23-32) that the quality of women Ph.D.s is at least equal to that of men in all objective respects. If the pool of women scientists were to be expanded to the size of the male pool, it is likely that the quality of the two would be identical. In principle, therefore, the pool of professional scientists can be expanded very considerably at no loss of quality; more important, the quality of that pool of constant size can also be raised by giving more encouragement to women than they have received so far. Intermediate options also exist, of course. It should be noted here that while various organizations that work with women graduate students (for example, Higher Education Resource Services) report such students do not currently perceive much active sex discrimination, in contrast to a decade ago, they also do not perceive the same encouragement as that given to male students.

The second conclusion is that the data presented here lend new emphasis to the necessity for adequate science and math education of girls at the pre-college level. The necessity for equal science and mathematics education of girls and boys throughout the school years cannot be overemphasized, and school science programs must insure that girls participate equally with boys. There are a number of programs aimed at female high school students, and their teachers and counselors, conducted on a local or regional basis, that have had positive results in encouraging interest in mathematics and providing career information. As an example, we note the Equals program in northern California which has worked with secondary school teachers to promote strategies for encouraging girls to continue with mathematics. The general topic of the quality of pre-college education in these fields has already aroused much concern on the part of the National Academy of Sciences and the National Science Foundation. We wish to record our urgent recommendation that potential new programs growing out of this concern be developed with careful regard to the equitable participation of female and minority students at all levels and from the beginning. In particular, Congressional authorization of such programs and potential state initiatives should include explicit support for such participation in the distribution of public funds. It is important to insure that every young person has the opportunity to assess his or her interest and talent for advanced study in mathematics and the sciences and for the pursuit of professions based on these fields, in order to insure that in the long term the best talent will be available in the advancement of science and in its application to national needs.

CHAPTER 2

THE SUPPLY OF WOMEN DOCTORATES

Since the turn of the century, the numbers of Ph.D.s awarded to women have increased continuously and, in the last decade, very steeply. This trend has been overshadowed in some periods, especially between World War II and the mid-1960s, by the steep rise in male Ph.D.s which resulted in large measure from the broadened educational participation of men made possible by the G.I. Bill and its successors after the Korean and Vietnam wars. While a few women were also entitled to such benefits, over 97 percent of the World War II veterans who received such aid were men. About one-quarter of all science Ph.D.s graduating in 1950 had received primary financial support from the G.I. Bill (Harmon, 1968); in 1981, 4-9 percent of new male science and engineering Ph.D.s and 0-.5 percent of women reported support from this source (Summary Reports, 1981). As a consequence, in part, of this disparate support pattern, the proportion of Ph.D.s granted to women in each decade reached an historic low in the 1950s and did not again match or exceed the previous high levels of the 1920s and 1930s until the last decade (Figure 2.1). The likelihood is considerable that the significantly lower availability of financial aid for women rather than the frequently cited baby boom was primarily responsible for their relative and temporary decline in the Ph.D. pool.

The numbers of women earning Ph.D.s in the sciences have increased steadily since 1970 while the numbers of men have declined (Table 2.1). Overall Ph.D. production peaked in 1973 and has stabilized in the last few years at around 18,000 per year. In the life and social sciences, total numbers of Ph.D.s continued to rise during most of the 1970s, which can be attributed entirely to the increase in female doctorates. The physical sciences have witnessed a substantial decline in doctoral production since 1971; the additional numbers of women graduates have not been large enough to offset the drop for male graduates from 5400 to 3600 annually.

Total production of engineering doctorates has also been declining since the early 1970s, although not as steeply as for the physical sciences. The number of women earning Ph.D.s in engineering is, despite a high rate of increase, still very low--only 100 women in 1981, compared with nearly 2400 men. However, a dramatic surge in female enrollments in engineering schools is now taking place.[1] The time lag before this increase can be expected to appear in the supply of women doctorates is 8-9 years, and thus at a date in the 1990s.[2]

In 1981 the percentages of women among new doctorates awarded were: physical sciences, 12 percent; engineering, 4 percent; life sciences, 26 percent; and social sciences, 36 percent. For all science and engineering fields combined, nearly 4,400 women earned Ph.D.s in 1981, representing 23 percent of the new recipients

Minority women in science

Of the 4,359 women receiving doctorates in science and engineering in 1981, 595 were minority group members (Table 2.1A), of whom Asians were the largest single group--313. Many of these women are presumably foreign citizens who completed graduate school with temporary visas and are returning to their home countries. (Fifty-six percent of recent Asian Ph.D.s are foreign citizens with temporary visas.[3]) Among Asian women doctorates a relatively high proportion graduated from physical sciences and engineering departments. Fifteen percent of the recent women Ph.D.s in the physical sciences were Asian.

Black women made up only 4 percent of all women receiving Ph.D.s in science and engineering in 1981. Their numbers have increased from 67 to 161 since 1974,[4] the first year for which data are available.

Institutional origins

A previous report of this Committee noted that women scientists receive their graduate training in roughly the same mix of institutions as men.[5] Table 2.2 repeats the analysis with information for the most recent doctorate recipients. In the majority of fields, the proportions

---

[1]The percent of women in freshmen engineering classes was: Fall 1973, 4.7 percent; Fall 1977, 11.1 percent; and Fall 1981, 15.8 percent. (Engineering Manpower Commission, unpublished data)

[2]Those entering college in Fall 1981 who complete the BA and go on to graduate school will receive the Ph.D. in approximately 1991.

[3]Syverson, 1981, p. 38.

[4]Doctorate Records File, National Research Council, unpublished data.

[5]Climbing the Academic Ladder, p. 31.

of men and women Ph.D.s who had graduated from highly rated departments[5a] are similar. In physics, women are somewhat less likely and in microbiology and psychology, somewhat more likely than men to have earned their doctorates from prestigious departments. The only field showing a substantial sex difference among 1976-1980 Ph.D.s is mathematics: 46 percent of the men, compared with 37 percent of the women received their graduate education from a highly rated department. It would be instructive to examine the distribution of recent men and women doctorate recipients for the 27 mathematics departments rated as "distinguished" or "strong" to determine whether this pattern is explained by a scarcity of women students in particular departments, or whether this statistical bias exists throughout the group.

The reader wishing more detailed data on both doctoral and baccalaureate origins of male and female Ph.D.s is directed to the report. A Century of Doctorates.[6] The appendices provide a unique listing of individual institutions ranked as Ph.D. producers within field and sex for various time periods.

Age at Ph.D.

Among recent doctorates in science and engineering, women typically complete their degrees at about the same age as men--31.0 for women and 30.3 for men. The median age at receipt of the doctorate varies widely according to field, with chemists and sociologists at the low and high ends of the range for both sexes (Figure 2.2). The only disciplines showing a marked gender difference are medical sciences and engineering.

In medical sciences, females are typically older than males receiving doctorates in the same year, with median ages of 33.7 and 30.3 respectively. An earlier report analyzing the graduates of the last 15 years noted that the standard deviations of age at Ph.D. are greater for women; the age spread for women in medical sciences was in fact greater than for any other science field, and was noticeably skewed at the upper age ranges,[6a] suggesting that for a portion of these women there are forces at work which have slowed their academic progress compared with their male counterparts. One might be quick to attribute this observation to marital or family constraints. However,

---

[5a]Based on reputational ratings of graduate faculty, as reported in Kenneth D. Roose and Charles J. Andersen, A Rating of Graduate Programs, American Council on Education, Washington, D.C., 1970.

[6]Harmon, 1978.

[6a]Ibid., p. 53.

FIGURE 2.1 Percent (and number) of science and engineering doctorates granted to women by field and decade, 1920-1979

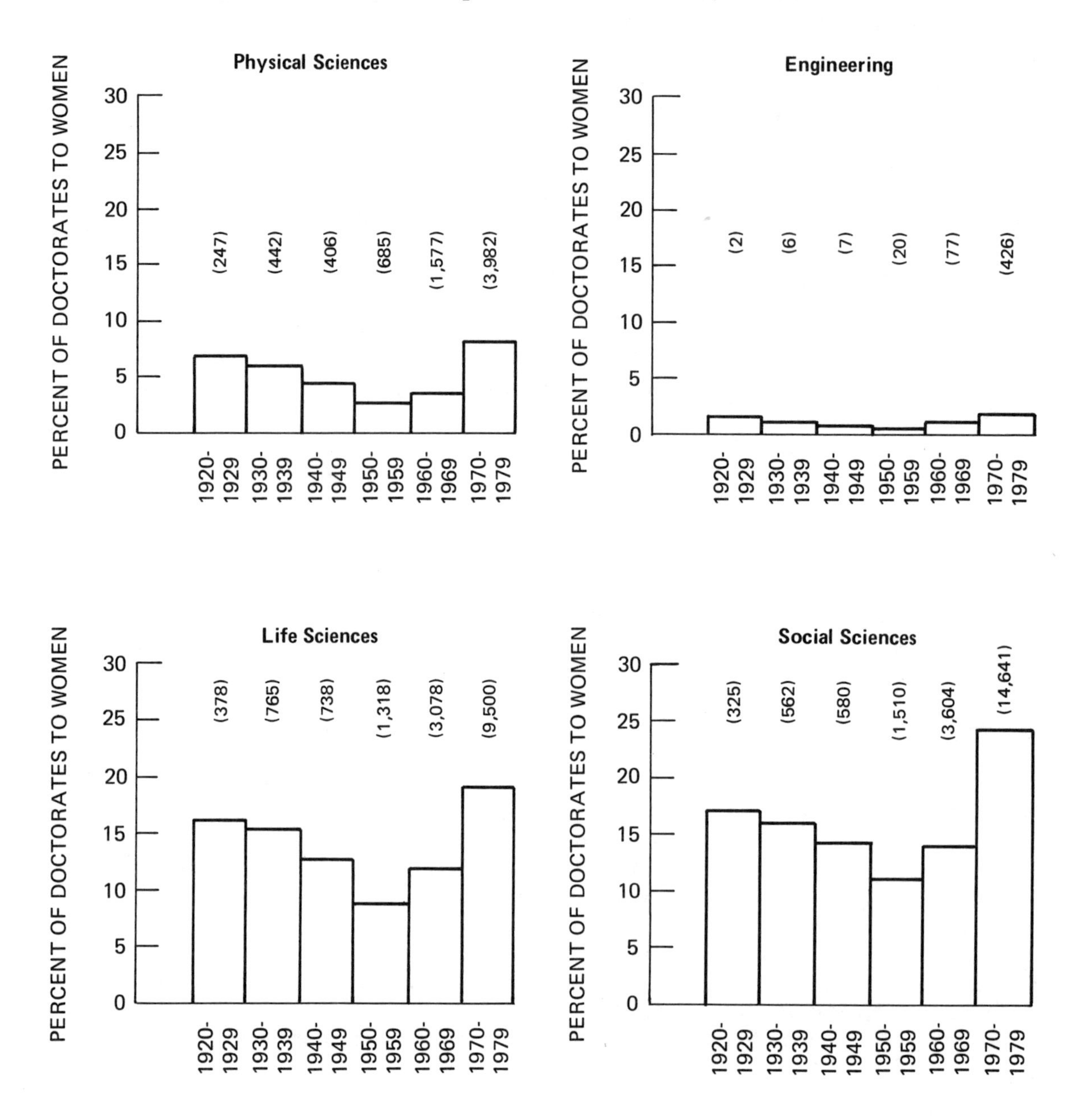

SOURCE: Doctorate Records File, National Research Council

TABLE 2.1 Number of science and engineering doctorates awarded by sex and field, 1970-1981

Year[a]	Total science & engrg.	Physical sciences	Engi-neering	Life sciences	Social sciences
			WOMEN		
1970	1,660	320	15	588	737
1971	1,995	341	15	715	924
1972	2,173	367	22	731	1,053
1973	2,542	382	46	868	1,246
1974	2,730	384	33	867	1,446
1975	3,005	403	52	950	1,600
1976	3,167	420	54	959	1,734
1977	3,298	430	74	957	1,837
1978	3,530	439	53	1,083	1,955
1979	3,861	496	62	1,194	2,109
1980	4,099	502	90	1,342	2,165
1981	4,359	502	99	1,443	2,315
			MEN		
1970	16,545	5,308	3,419	3,989	3,829
1971	17,506	5,398	3,483	4,360	4,265
1972	17,431	5,171	3,481	4,221	4,558
1973	17,079	4,929	3,318	4,140	4,692
1974	16,400	4,592	3,114	3,967	4,727
1975	16,069	4,454	2,950	3,955	4,710
1976	15,646	4,089	2,780	3,921	4,856
1977	15,025	3,949	2,569	3,816	4,691
1978	14,442	3,754	2,370	3,808	4,510
1979	14,401	3,803	2,428	3,887	4,283
1980	14,072	3,612	2,389	3,983	4,088
1981	14,303	3,666	2,429	4,018	4,190

[a]Fiscal year is used throughout the tables in this chapter. Fiscal year 1980, for example, represents the period July 1, 1979 through June 30, 1980.

SOURCE: Syverson, 1981, p. 4.

TABLE 2.1A Number of science and engineering doctorates awarded in 1981 by sex, racial-ethnic group, and field[a]

Racial-ethnic group	Total, science & engr.	Physical sciences	Engi-neering	Life sciences	Social sciences
			WOMEN		
TOTAL	4,359	502	99	1,443	2,315
White	3,499	371	58	1,146	1,924
Asian	313	77	31	126	79
Black	161	11	--	42	108
American Indian	8	--	--	2	6
Puerto Rican	19	4	--	5	10
Mexican	20	1	--	3	16
Other Hispanic	74	13	3	25	33
Other & unknown	265	25	7	94	139
			MEN		
TOTAL	14,303	3,666	2,429	4,018	4,190
White	10,388	2,676	1,341	3,079	3,292
Asian	1,917	552	773	357	235
Black	435	58	59	135	183
American Indian	23	2	4	9	8
Puerto Rican	26	10	5	5	6
Mexican	98	13	18	31	36
Other Hispanic	337	59	69	136	73
Other & Unknown	1,079	296	160	266	357

[a]Non-U.S. citizens with temporary visas are included.

SOURCE: Doctorate Records File, National Research Council.

TABLE 2.2 Percent of science and engineering doctorates awarded to men and women from highly rated departments,[a] for selected fields, 1970-1980

Year and field of doctorate	Total doctorates awarded, all departments		Percent from highly rated departments[a]	
	Women	Men	Women	Men
1970-1975				
Mathematics	612	6,722	42	46*
Physics	282	8,225	39	46*
Chemistry	1,110	10,786	45	45
Biochemistry	748	2,932	42	44
Microbiology	534	1,797	40	29*
Psychology	3,974	10,147	39	32*
Anthropology	556	1,251	55	52
Sociology	896	2,759	47	40*
Economics	383	4,876	49	39*
1976-1980				
Mathematics	569	3,719	37	46*
Physics	263	4,640	41	48*
Chemistry	1,038	6,805	44	45
Biochemistry	768	2,340	39	42
Microbiology	512	1,220	34	30*
Psychology	5,731	9,386	29	25*
Anthropology	798	1,167	43	41
Sociology	1,147	2,157	36	38
Economics	467	3,628	39	39

[a]Based on ratings of "distinguished" or "strong" graduate faculty, as reported in Kenneth D. Roose and Charles J. Andersen, A Rating of Graduate Programs, American Council on Education, Washington, D.C., 1970.

*Sex difference in percent from highly rated departments is statistically significant at .05 level.

SOURCE: Doctorate Records File, National Research Council

FIGURE 2.2 Median age at Ph.D. by field and sex, 1981 science and engineering doctorates

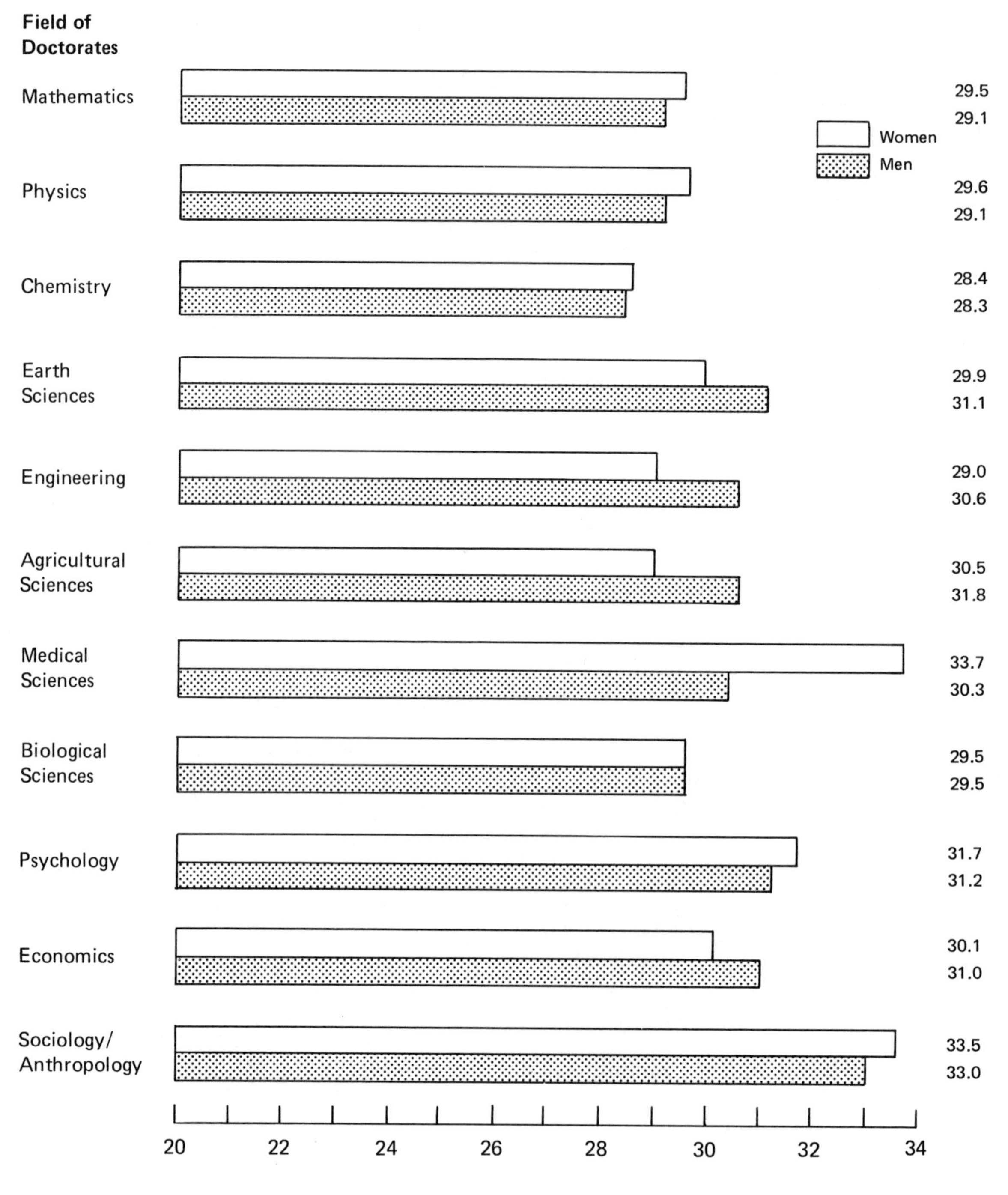

SOURCE: Syverson, 1982, pp. 53-56.

only 47 percent of the new women Ph.D.s in medical sciences were married, a somewhat lower proportion than in chemistry, where the female graduates are comparatively young at the time of Ph.D.

It is probable that this large age differential arises from different field distributions by sex within the medical sciences (see also p. 2.9 and p. 3.2); more than a quarter of the women in these fields obtain Ph.D.s in nursing, and are likely to have spent some period of time in professional practice before undertaking doctoral work. Such a conclusion is also supported by the fact that 72 percent of the women but only 55 percent of the men have taken a master's degree.[7]

In engineering the sex pattern is reversed: women typically receive their Ph.D. at a considerably younger age than do men. Here an explanation for a sex difference is not apparent, especially since the women are more likely to have switched into engineering from other baccalaureate fields, a pattern which would be expected to have prolonged their graduate study.

The significance of sex differences in age at receipt of the Ph.D. is not entirely clear in any case; traditionally, earning the doctorate at an early age has been seen as an indication of high interest, motivation, and ability. However, even cursory inspection of the median BA-to-Ph.D time lapse by field and sex for different institutions (see Harmon, Table 42) shows very wide variations among the leading universities; the difference in median time lapse between two institutions of equal eminence is frequently greater than the sex difference in either one. Such variations may be accounted for by differing policies among institutions with respect to residence requirements, financial support, TA duties, and other factors. Thus sex differences in age at receipt of the doctorate can be influenced markedly by differential sex distributions among specific departments even within given categories of quality.

Graduate school support patterns

Support patterns are strikingly similar for recent male and female graduate students in science and engineering departments. Within fields, there is very little difference in the percentages of men and women who held research assistantships or teaching assistantships. The most pronounced change in financial aid over recent years is the reduced availability of fellowships, which has been found to have affected both sexes about evenly.[8]

Medical sciences is somewhat of an exception. The women doctorates are twice as likely as the men (26 percent versus 12 percent) to report

---

[7]Syverson, 1981, pp. 32-35.

[8]Ahern and Scott, 1981, pp. 6-7.

personal funds (i.e., own earnings, spouse's earning, or family contributions) as their primary source of financial support during graduate school.[9]

We noted in connection with the large age differential by sex in this field (p. 2.3) that this probably derives from the large number of women Ph.D.s in nursing. The large differential in financial support may arise from the same reason; if the total support available for research degrees in nursing is significantly less than in other medical fields, then women in nursing are disproportionately burdened by having to provide more of their own support.

Predoctoral employment

Male and female doctorates are nearly equally likely to be teaching full-time prior to the doctorate,[10] but only a small proportion were so employed--about 4 percent in physical sciences, 6 percent in life sciences, and 9 percent in engineering. In the social sciences, however, as many as 23 percent of the women and 31 percent of the men reported full-time teaching prior to receiving their degrees.

TABLE 2.3 Percent of 1977 and 1981 science and engineering Ph.D.s who were married at receipt of the doctorate, by field and sex

Field of doctorate	1977 % Married		1981 % Married	
	Women	Men	Women	Men
Mathematics	57.0	58.8	41.1	51.3
Physics	68.8	60.7	61.6	49.0
Chemistry	52.2	62.6	51.9	52.9
Earth sciences	52.5	72.3	44.6	61.0
Engineering	66.2	67.7	55.6	61.8
Agriculture	42.9	79.4	57.1	75.4
Medical sciences	48.5	72.1	47.4	65.1
Biological sciences	50.1	68.2	45.9	58.3
Psychology	50.5	63.0	47.4	58.8
Social sciences	51.0	70.0	49.6	63.2

SOURCE: National Research Council, Summary Report: Doctorate Recipients from U.S. Universities, 1977 and 1981 reports in the series.

[9] Summary Report 1981: Doctorate Recipients from U.S. Universities, Text Table D. (in press)

[10] Survey of Doctorate Recipients, National Research Council, unpublished data.

Marital status

Generally speaking, women Ph.D.s are less likely than their male counterparts to be married at the time of receiving their degrees (Table 2.3). For 1981 doctorate recipients, the proportion of married women is below 50 percent in most science fields. Marriage is less common among new Ph.D.s than it was just 4 years ago, but the decreases are more dramatic for men than for women.

Plans after the Ph.D.

The Survey of Earned Doctorates obtains responses from individual doctorate recipients on their plans for employment or further study immediately following graduation, including the employment sector they plan to enter, and whether they already have a definite job, or are still seeking an appointment. The percent of new Ph.D.s with definite jobs at receipt of the degree has been used as a barometer for the state of the doctoral labor market, as these data are available by field on an annual basis. The data also indicate the comparative status of men and women graduates in obtaining employment.

Table 2.4 shows that the new women Ph.D.s are somewhat less likely than their male counterparts to have a definite appointment at the time of receiving the degree. For example, in the biological sciences--a field with a substantial female component--38 percent of the women were still seeking a position compared with 29 percent of the men. Chemistry, with women representing one-sixth of the new Ph.D.s, shows a similar differential: 25 percent of the women but only 16 percent of the men reported that they were still looking for employment. Physics is the single field in which this pattern is reversed.

A longer waiting time in obtaining a first appointment may continue to be a disadvantage in the early career years. It has been shown in the case of humanities Ph.D.s that those who are seeking but do not have a position at the time of graduation are more likely to be unemployed 1-2 years later (Hornig, unpublished data).

The academic sector continues to be the largest employer of doctorate recipients. The percent of new Ph.D.s planning to have an academic position immediately after graduation ranges from 9 percent in chemistry to 56 percent in mathematics and may differ for men and women (Table 2.5). The reader is reminded that the sizeable group of doctorates with "other plans" shown in Table 2.5 includes those taking postdoctoral fellowships, associateships, traineeships, etc. (The trend toward increasing numbers of postdocs is discussed in detail in Chapter 3.) The group with "other plans" also includes those going into government employment (8 percent of the physics Ph.D.s and 10 percent of the engineers).

TABLE 2.4 Employment prospects at time of receipt of the doctorate, by field and sex, 1980 science and engineering Ph.D.s

	Total doctorates		Number planning employment[a]		% With definite job		% Still seeking job	
	Women	Men	Women	Men	Women	Men	Women	Men
Mathematics	95	650	86	508	72%	80% *	28%	20%
Computer sciences	21	197	20	166	80	79	20	21
Physics	67	918	29	410	86	76 *	14	24
Chemistry	255	1283	127	702	75	84 *	25	16
Earth sciences	64	564	37	372	76	82	24	18
Engineering	90	2389	77	1903	79	81	21	19
Agricultural sciences	109	963	79	756	66	80 *	34	20
Medical sciences	278	564	170	263	78	83	22	17
Biological sciences	955	2456	280	775	62	71 *	38	29
Psychology	1310	1788	993	1362	68	73 *	32	27
Economics	103	664	94	610	80	85	20	15
Sociology & anthropology	405	566	331	476	64	68	36	32
Other social sciences	347	1070	301	928	68	76 *	32	24

[a] Includes those Ph.D.s who at the time of receiving the doctorate, planned to be employed as opposed to those planning to take a postdoctoral fellowship, traineeship, etc. The numbers and characteristics of those planning post-doctoral study are discussed in Chapter 3.

*Sex difference in percent with definite job is statistically significant at .05 level.

SOURCE: Doctorate Records File, National Research Council

TABLE 2.5 Percent of 1980 science and engineering doctorates planning academic and industrial employment following receipt of the Ph.D.

	% planning acad. employ.	% planning indust. employ.	% other plans
WOMEN			
Mathematics	65%	18%	17%
Computer sciences	52	29	19
Physics	10	25	65
Chemistry	14	33	53
Earth sciences	28	17	55
Engineering	28	42	30
Agricultural sciences	33	16	51
Medical sciences	42	8	50
Biological sciences	17	5	78
Psychology	33	11	56
Economics	52	15	33
Sociology and anthropology	56	4	40
Other social sciences	58	14	28
MEN			
Mathematics	55	15	30
Computer sciences	39	38	23
Physics	12	21	67
Chemistry	8	41	51
Earth sciences	23	24	53
Engineering	23	42	35
Agricultural sciences	44	13	43
Medical sciences	21	10	69
Biological sciences	17	7	76
Psychology	28	12	60
Economics	56	9	35
Sociology and anthropology	59	5	36
Other social sciences	63	9	28

Industrial employment is, not surprisingly, highest in engineering, computer sciences and chemistry. About 40 percent of the new Ph.D.s in these departments report that they have or are looking for a position in business or industry. Characteristics of male and female Ph.D.s employed in industry are described in Chapter 5.

Labor force participation

Approximately 95 percent of the women and 99 percent of the men who earned Ph.D.s in science and engineering in the 1970s were in the labor force as of 1981.[10a] Thus, very few women scientists--only one in 20 of the recent Ph.D. cohorts--chose not to work. Their attachment to the labor force is higher than is commonly believed--even among women in the childbearing age groups.[11]

Women now make up 12 percent of the total U.S. doctoral work force in science and engineering (Table 2.6). They account for only 1 percent of the supply of all engineering Ph.D.s, 3 percent of the doctoral physicists, 8 percent of Ph.D. chemists, 18 percent of bioscience Ph.D.s, and as many as 27 percent of the supply of Ph.D. psychologists. In 1981 there were nearly 41,000 doctoral women scientists and engineers in the U.S. work force.

The majority of the women scientists are working full-time. Part-time employment is reported by 11 percent of the women compared with 2 percent of the men, across all employment sectors. In psychology, one in six of the women Ph.D.s hold part-time jobs while in mathematics and engineering, only one in 13 do so. About 20 percent of the women who were working part-time reported that they were seeking a full-time position.[12]

Women scientists are more likely than their male colleagues to be unemployed involuntarily although for both sexes the numbers are relatively low. The 1981 unemployment rates, calculated separately by field, remain at 1 percent or below for men, regardless of field, and at about 2 percent for women in most fields, although unemployment levels reach 3-4 percent for female Ph.D.s in the biomedical areas and the social sciences.[13]

[10a]National Research Council, Survey of Doctorate Recipients, unpublished data.

[11]The median age of the 1976-1980 women Ph.D.s at the time of the survey was 33; 96 percent reported that they were employed or seeking employment.

[12]National Research Council, 1981 Profile: Science, Engineering, and Humanities Doctorates in the U.S., Table 1.10. (forthcoming)

[13]Ibid., The unemployment rate is the ratio of the number who were unemployed and seeking work to the number in the labor force.

TABLE 2.6 Number and percent of women doctoral scientists and engineers in the labor force by field, 1981

	BY FIELD OF DOCTORATE:		BY FIELD OF EMPLOYMENT:	
	Number of women	Women as % of total	Number of women	Women as % of total
All science & engineering fields	40,852	12.0	35,569	11.5
Mathematics	1,482	8.1	1,165	8.4
Computer sciences	178	8.2	571	6.9
Physics	832	2.9	547	2.9
Chemistry	3,769	7.8	2,758	7.2
Earth sciences	588	4.9	800	5.2
Engineering	504	1.0	716	1.4
Agricultural sciences	453	3.0	366	2.5
Medical sciences	1,789	17.0	3,582	17.6
Biological sciences	10,392	18.2	8,309	17.8
Psychology	12,257	26.9	10,437	27.1
Economics	1,153	8.4	900	8.3
Other social sciences	7,455	19.5	5,418	18.1

SOURCE: Survey of Doctorate Recipients, National Research Council

CHAPTER 3

POSTDOCTORAL TRAINING

Recent attention has been given to the status of the growing postdoctoral population in science departments.[1] As junior faculty appointments became more scarce over the past decade, growing numbers of young Ph.D.s in the biosciences and physics applied for and accepted postdoctoral appointments--usually research associateships--(Figure 3.1),

FIGURE 3.1 Percent of new Ph.D.s planning postdocs, 1972-1980

SOURCE: Doctorate Records File, National Research Council

[1]National Research Council, Postdoctoral Appointments and Disappointments, (National Academy Press), 1981.

some presumably to hold them over until a faculty position became available. The pay is typically low; there are in fact a few documented cases of unpaid positions.[2] In this chapter, we will examine recent patterns for female and male postdocs, including their numbers, reasons for taking a postdoctoral appointment, the institutions which sponsor them, the average stipend levels, and the relationship of marital status to some of these factors.

In the fields where postdoctoral appointments are a common pattern--physics, chemistry, the earth sciences, agricultural and biosciences--similar proportions of recent men and women Ph.D.s reported that they planned to take a postdoc (Table 3.1). In medical sciences also, a substantial fraction of the 1981 Ph.D.s plan postdoctoral study but for women it is only one-third compared with one-half of the men. This is perhaps explained by the different field distributions of men and women within the medical sciences group, specifically with respect to nursing. In 1980, 74 out of 278 women received doctorates in nursing versus 2 out of 564 men.

For both sexes, there has been very little change since 1977 in the fraction taking postdocs except for a drop in chemistry.

Reason for taking a postdoctoral appointment

Both men and women report that their primary reason for taking a postdoc is to gain additional research experience in their field of Ph.D. (Table 3.2). Other reasons cited include wanting to work with a particular scientist or research group (about 15-20 percent) or the desire to switch into a different field (10-20 percent). In medical sciences, women more often than men reported switching fields. Also, in chemistry the men are significantly more likely to state that they are taking a postdoc to work with a particular scientist or research group. An apposite finding from other studies (e.g., Feldman, 1974; Perucci, 1975) is that chemistry departments have tended to provide inadequate opportunities for professional socialization to women students.

A surprisingly small fraction of men or women report that inability to obtain employment was the major reason for taking a postdoc. About one-fifth of the psychology Ph.D.s reported "other reasons." Further examination of the data reveals that these are clinical and counselling psychologists who are likely taking a postdoctoral internship.

In most fields and for both sexes, about three-fourths of the Ph.D.s were able to secure early appointments with one-fourth still searching (Table 3.3). By far the largest numbers of postdocs are in

[2] "Slave Labor on Campus: The Unpaid Postdoc," Science, Vol. 216, May 14, 1982, pp. 714-715.

TABLE 3.1 Number and percent of 1977 and 1981 science and engineering doctorates planning postdoctoral appointments by field and sex

Field of doctorate	Women: Total doctorates	Women: Number Planning postdoc	Women: % Planning postdoc	Men: Total doctorates	Men: Number planning postdoc	Men: % Planning postdoc
1977						
Mathematics	128	12	9%	831	91	11%
Physics	64	33	52	1,085	499	46
Chemistry	180	92	51	1,390	639	46
Earth sciences	59	22	37	632	158	25
Engineering	74	12	16	2,567	385	15
Agricultural sci.	63	13	21	861	112	13
Medical sciences	165	61	37	506	202	40
Biological sci.	729	452	62	2,443	1,417	58
Psychology	1,081	162	15	1,879	301	16
Social sciences	749	59	7	2,795	152	5
1981						
Mathematics	112	9	8%	616	105	17%
Physics	73	34	47	942	424	45
Chemistry	235	87	37	1,376	537	39
Earth sciences	56	14	25	526	158	30
Engineering	99	13	13	2,429	316	13
Agricultural sci.	147	25	17	1,003	130	13
Medical sci.	310	99	32	604	284	47
Biological sci.	986	651	66	2,411	1,543	64
Psychology	1,472	265	18	1,885	320	17
Social sciences	843	58	6	2,305	164	7

SOURCE: Syverson, 1978, pp. 22-25; 1982, pp. 53-56.

TABLE 3.2 Reason for taking a postdoctoral appointment,[a] 1980 doctorates in selected science fields by sex

	For additional research experience in Ph.D. field	To work with particular scientist or research group	To switch into different field	Could not obtain employ.	Other reason
Physics					
Women (n = 18)	61%	11%	22%	0%	6%
Men (n = 268)	68	15	10	6	2
Chemistry					
Women (n = 94)	64	13	15	4	4
Men (n = 398)	52	19	16	11	2
Medical sciences					
Women (n = 65)	60	15	17	6	2
Men (n = 192)	60	17	9	5	9
Biological sciences					
Women (n = 456)	54	20	16	5	5
Men (n = 1,126)	57	22	12	6	3
Psychology					
Women (n = 157)	41	19	8	8	24
Men (n = 200)	43	19	9	11	20

SOURCE: Doctorate Records File, National Research Council

[a] Respondents were asked to check the primary reason.

TABLE 3.3 Status of postdoctoral appointment at time of receipt of Ph.D., by field and sex, 1980 science and engineering Ph.D.s

Field of doctorate[a]	Number planning postdoctoral		% With definite appointment		% Still seeking appointment	
	Women	Men	Women	Men	Women	Men
Physics	34	443	70%	80%	30%	20%
Chemistry	115	524	76	77	24	23
Earth sciences	22	165	77	81	23	19
Agricultural sciences	25	149	72	59	28	41
Medical sciences	93	273	81	84	19	16
Biological sciences	621	1,559	80	81	20	19
Psychology	223	294	72	75	28	25
Social sciences	80	147	62	63	38	37

[a]Shown are fields with more than 20 women planning postdoctoral appointments.

SOURCE: Doctorate Records File, National Research Council

the biosciences: approximately 500 women and 1,250 men had definite awards by the time they received their Ph.D. Although relatively few social scientists--male or female--seek postdoctoral fellowships, of those who did in 1980, only 60 percent reported that they had an appointment in hand upon graduation.

Marital status and postdoctoral patterns

About one-fifth of all postdoctorals have held long-term appointments--for more than 36 months (Table 3.4). A "holding" pattern is evident in both the physical and life sciences. Among life scientists married women are more likely than single women to be in this category. They also report that their postdoc status was prolonged because of difficulty in finding employment. For men, the pattern is reversed, with single men more likely to hold long-term appointments.

Marital status plays a role in the postdoc decision in yet another way. In a survey of 1978 Ph.D.s who were postdocs, as many as 70 percent of the married women cited geographic limitations as an "important factor" in their taking a postdoctoral appointment. Geographic constraints were also a deciding factor for 33 percent of the single women and about 25 percent of the men postdoctorals (Table 3.5).

Host institutions

Similar proportions of men and women postdocs with new awards in 1980 were accepted at major research universities. Table 3.6 shows the distribution of new postdoctoral appointees by type of institution and sex. It should be noted that these numbers are based on those reporting definite acceptance by the time of receiving the doctorate; the total number of 1980 Ph.D.s who received appointments for the following year may be higher.

In chemistry, the top 25 institutions accounted for about 40 percent of the new postdocs awarded, for both men and women. Because of the comparatively small number of female postdocs in a single year, however, this translates to 29 women with new awards in chemistry or roughly one per top-25 department. In the biosciences, the leading research universities were the destination for about 25 percent of the postdocs, with no difference by sex.

TABLE 3.4 Percent of 1972 Ph.D. recipients who held long-term (>36 months) postdoctoral appointments by sex and marital status

	Women			Men		
	Total	Single	Married	Total	Single	Married
All science/engr. fields						
Total taking postdoc	501	230	271	3,750	1,033	2,717
Prolonged postdoc because of difficulty in finding employment	30%	25%	34%	28%	35%	26%
Held postdoc appt. >36 months	23	21	24	18	24	15
Engineering, mathematics, physical sciences						
Total taking postdoc	83	35	48	1,941	594	1,347
Prolonged postdoc because of difficulty in finding employment	(43)%	(43)%	(44)%	32%	36%	31%
Held postdoc appt. >36 months	20	(20)	(21)	16	20	13
Life sciences						
Total taking postdoc	328	152	176	1,368	354	1,014
Prolonged postdoc because of difficulty in finding employment	31%	25%	36%	28%	37%	25%
Held postdoc appt. >36 months	29	25	32	24	33	21
Social sciences						
Total taking postdoc	90	43	47	441	85	356
Prolonged postdoc because of difficulty in finding employment	14%	(14)%	(15)%	11%	21%	9%
Held postdoc appt. >36 months	3	7	0	6	7	6

NOTE: Percentage estimates reported in this table are derived from a sample and are subject to an absolute sampling error of less than 5 percentage points, unless otherwise indicated. Estimates with sampling errors of 5 or more percentage points are reported in parentheses.

SOURCE: National Research Council, 1981, p. 152.

TABLE 3.5 Geographic limitations as a factor in taking a postdoctoral appointment, by sex and marital status, 1978 science and engineering Ph.D.s

	Women		Men	
	Single	Married	Single	Married
All science/engineering fields				
Total taking postdoc	463	436	1,465	1,742
Geographic limitations				
Important factor	33%	70%	22%	26%
Incidental factor	23	8	25	25
Not a factor	44	22	52	50
Engineering, mathematics, physical sciences				
Total taking postdoc	59	52	704	605
Geographic limitations				
Important factor	24%	(60)%	22%	26%
Incidental factor	17	4	29	23
Not a factor	59	(36)	49	51
Life sciences				
Total taking postdoc	234	237	554	954
Geographic limitations				
Important factor	26%	71%	16%	25%
Incidental factor	22	8	25	25
Not a factor	52	21	58	50
Social sciences				
Total taking postdoc	170	147	207	183
Geographic limitations				
Important factor	(44)%	(73)%	(40)%	28%
Incidental factor	26	10	13	27
Not a factor	29	18	(47)	44

NOTE: Percentage estimates reported in this table are derived from a sample survey and are subject to an absolute sampling error of less than 5 percentage points, unless otherwise indicated. Estimates with sampling errors of 5 or more percentage points are reported in parentheses.

SOURCE: National Research Council, 1981, p. 151.

TABLE 3.6 Host institutions[a] for 1980 Ph.D.s with definite postdoctoral appointments, by sex and field

	WOMEN				MEN			
	Total no. postdocs, excluding medical schools	%			Total no. postdocs, excluding medical schools	%		
		Top 25	Second 25	Other inst.		Top 25	Second 25	Other inst.
Physics	12	50	17	33	216	41	13	47
Chemistry	74	39	16	45	322	40	15	45
Biological sci.	389	25	12	64	1,026	24	12	64
Psychology	92	33	11	57	150	25	9	66

[a]Institutions are categorized by federal R&D expenditures. For a listing of the institutions, see Appendix C.

SOURCE: Doctorate Records File, National Research Council

Postdoctoral stipends

As of 1981, women postdocs were paid roughly the same as men, judging by median stipends in chemistry and biosciences (Figure 3.2). This contrasts with the situation about 12 years ago when women postdoctorals were reported to be earning an average of about $1400 less than men and just 4 years ago when the pay differential was estimated at $800.[3] As noted in the Committee's earlier report, equity at this level is important in that one might expect it would lead to comparable salaries in subsequent employment. Whether the latter statement has been realized will be examined in the following chapter.

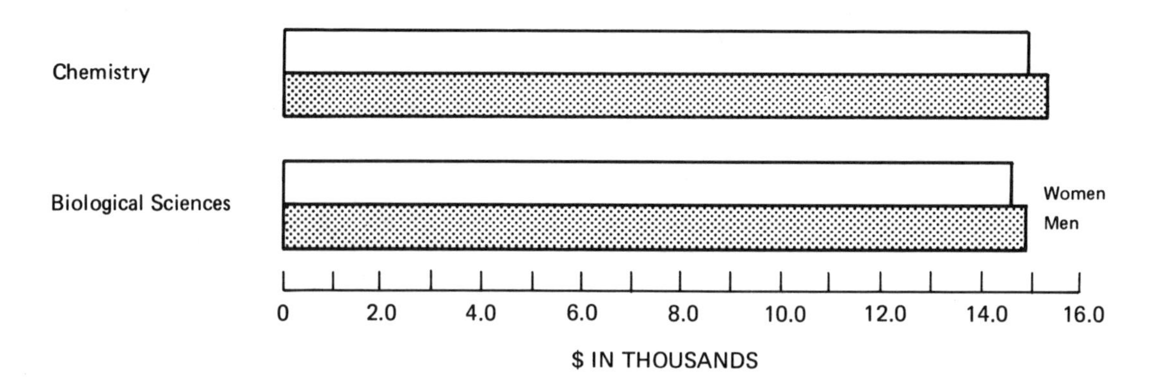

FIGURE 3.2 Postdoctoral stipends[a] in two fields[b] by sex, 1981

[a]Median annual stipends. Stipends reported for a 9-10 month period have been adjusted to a full-year equivalent. Includes both first-year and renewed appointments.

[b]These two fields were selected because they have substantial numbers of female postdocs.

SOURCE: Survey of Doctorate Recipients, National Research Council

[3]Invisible University (1969) and Climbing the Academic Ladder.

CHAPTER 4

WOMEN SCIENTISTS AND ENGINEERS IN ACADEME

Rates of faculty growth have slowed considerably in the past decade. Between 1977 and 1981, the number of full-time faculty in all fields increased by only 1 percent a year, and for the eight preceding years, by 3 percent a year compared with 10 percent annual growth in the 1960s.[1] Even so, there has been a substantial increase in the number of women on science and engineering faculties, an increase of 3,200 in a recent 4-year period.[2] How this larger, somewhat younger population of women scientists and engineers fared in terms of ladder appointments, tenure, rank and salaries, will be examined in this section.

Academic vs. nonacademic employment

The majority of doctoral women scientists and engineers (59 percent) continue to be employed in colleges and universities but the academic fraction has dropped since 1977 when it was 65 percent. As a group, women are still more likely than men to be in academe (Figure 4.1); when the comparison of men and women is controlled by field, however, the sex difference nearly disappears.[3] Doctoral employment in business and industry has increased for both women and men since 1977. Here again, the much lower incidence of industrial employment for women scientists is largely but not entirely a function of the different field distributions, with very small proportions of women in engineering and physics.

Overall, women scientists are twice as likely as men to be employed in state and local government and hospitals or clinics, and much more likely to be in nonprofit organizations, partly due to their different field distributions. It is probable that some scientists, at least in

---

[1]National Center for Education Statistics, Digest of Education Statistics, 1982.

[2]See Table 4.2, page 4.6.

[3]National Research Council, Science, Engineering, and Humanities Doctorates: 1981 Profile (in press).

FIGURE 4.1 Percent distribution of doctoral scientists and engineers by employment sector and sex, 1981

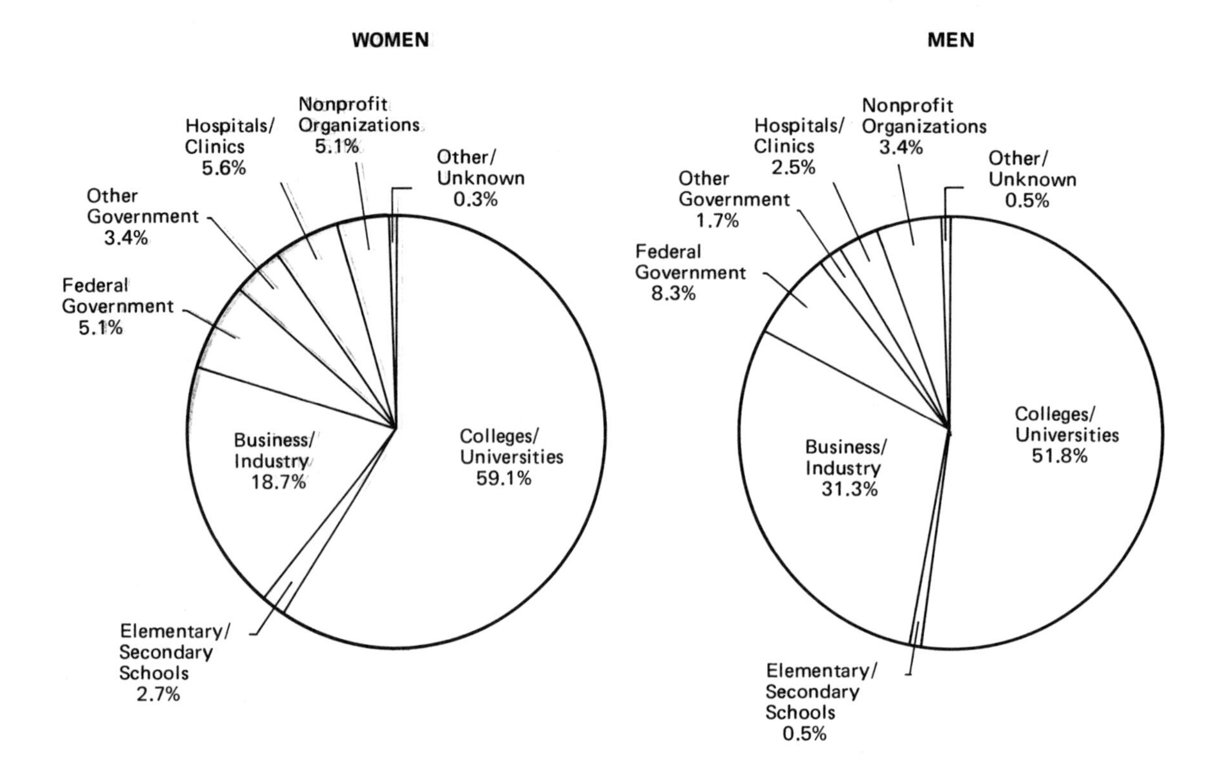

SOURCE: Survey of Doctorate Recipients, National Research Council

the physical sciences, took such positions only as a last resort in preference to being unemployed, and are then frequently unable to contribute to the development of their fields in any substantial way.

Full-time and part-time employment

More than 90 percent of doctoral women in academe are full-time employed (Table 4.1). The incidence of part-time employment among academically employed women scientists varies widely by field: in physics, chemistry, earth sciences, and biosciences one in seven women Ph.D.s hold part-time positions while in several other fields, the rate is only one in 20. Based on the total number of part-time faculty in all fields, women are a minority, comprising 40 percent of the part-timers at universities, 41 percent at 4-year institutions, and 37 percent at 2-year institutions,[4] but these proportions all exceed women's representation in the total doctorate pool considerably.

[4] *On Campus with Women*, Project on the Status and Education of Women, Association of American Colleges, No. 31, Summer 1981.

TABLE 4.1 Full-time and part-time employment of doctoral scientists and engineers in academe[a] by field and sex, 1981

Employment field	Number employed[b] Women	Men	% Full-time Women	Men	% Part-time Women	Men
All science and engineering fields	11,902	96,697	90.3	98.6	9.8	1.3
Mathematics	623	7,845	88.9	99.3	11.1	0.7
Computer sciences	142	1,982	93.6	96.2	6.3	3.8
Physics	203	7,011	85.2	99.0	14.8	1.1
Chemistry	702	7,453	86.2	97.7	13.8	2.3
Earth sciences	283	4,746	85.9	97.9	14.1	2.1
Engineering	161	13,961	91.3	99.1	8.7	0.9
Agricultural sciences	178	7,728	94.3	99.3	5.6	0.6
Medical sciences	809	2,716	97.7	98.5	2.3	1.4
Biological sciences	2,590	13,447	86.4	98.9	13.6	1.0
Psychology	2,615	9,042	90.2	97.9	9.8	2.1
Economics	374	5,287	96.2	99.3	3.7	0.7
Sociology/anthropology	1,478	4,508	93.0	98.0	7.0	1.9
Other social sciences	1,744	10,971	91.3	98.5	8.8	1.5

[a]Includes 2-year and 4-year colleges and universities.

[b]Excludes postdoctorals.

SOURCE: Survey of Doctorate Recipients, National Research Council.

Numbers of women faculty[5]

As of 1981, there were approximately 13,500 doctoral women on U.S. science and engineering faculties, accounting for 10.9 percent of the total. Their representation is up from 9.3 percent in 1977 (Table 4.2).

In terms of faculty growth, women represented 24 percent of the increase over all science and engineering departments between 1977 and 1981. The major research universities as a group had the largest change in percentage of women, despite the fact that the "other" institutions showed more expansion during this period. The increase in the major research institutions, however, occurred on a very much smaller base, and the percentages of women faculty by field remain well below the percentages of women scientists in the relevant fields and Ph.D. cohorts, except among assistant professors (see below).

One of the issues of concern to this Committee is the effect that such increases may have on the climate within departments and particularly on the important function of presenting to both women and men students an image of women as effective scientists. In that regard, it is instructive to note that a total increase of 865 women science faculty divided among a minimum of 13 departments (see the fields listed in Table 4.1--a more realistic number is probably 15-18) in the 50 top universities results in adding an average of 1.3 female faculty in four years to each department. It is hardly necessary to stress that from the point of view of either students or male faculty, this is not yet a startling change. In doctorate-granting departments of both physics and chemistry, many institutions still have no women faculty at all (APS, 1982; ACS, 1980).

Sex distribution of faculty appointments

Approximately 50 percent of all males in science and engineering departments were full professors in 1981 (Figure 4.2), with the major research universities more "top-heavy" than other institutions. And although there were 3,000 doctoral women scientists employed in the leading institutions, only 10 percent of the women were full professors; 43 percent were in off-ladder positions or are postdoctoral appointees.

[5]Throughout this chapter, faculty statistics include four-year colleges and universities only. The 1977 data presented here may differ from 1977 numbers in the Committee's first report due to the fact that medical schools were formerly included in the totals.

TABLE 4.2 Increase in doctoral scientists and engineers in faculty[a] positions by R&D expenditures of institution[b] and sex, 1977-1981

		Total science & engineering faculty	Number of women	Women as % of 1977-1981 increase
Total all inst.[c]	1981	123,660	13,471	
	1977	110,201	10,231	
	4-yr. growth	13,459	3,240	24.1%
Total first 50 inst.	1981	31,328	2,754	
	1977	28,257	1,889	
	4-yr. growth	3,071	865	28.2%
Top 25 inst.	1981	17,369	1,718	
	1977	15,401	1,160	
	4-yr. growth	1,968	558	28.4%
Second 25 inst.	1981	13,959	1,036	
	1977	12,856	729	
	4-yr. growth	1,103	307	27.8%
Other inst.	1981	92,332	10,717	
	1977	81,944	8,342	
	4-yr. growth	10,388	2,375	22.9%

[a]Faculty includes professor, associate professor, and assistant professor ranks.

[b]See Appendix for a description of ranking of institutions by federal R&D expenditures.

[c]Includes 2-year and 4-year colleges and universities. Excluded are those employed at medical schools and university-administered federal laboratories.

SOURCE: Survey of Doctorate Recipients, National Research Council.

FIGURE 4.2 Faculty rank distribution of doctoral scientists and engineers by R&D expenditures of institution* and sex, 1981

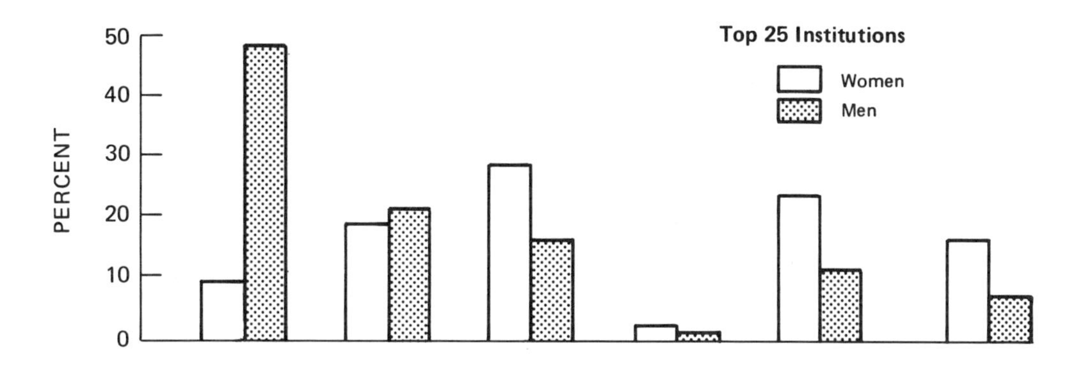

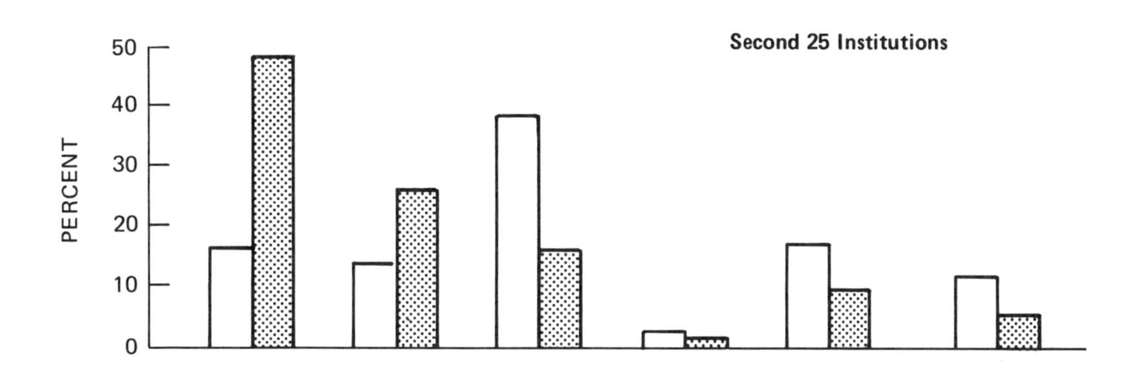

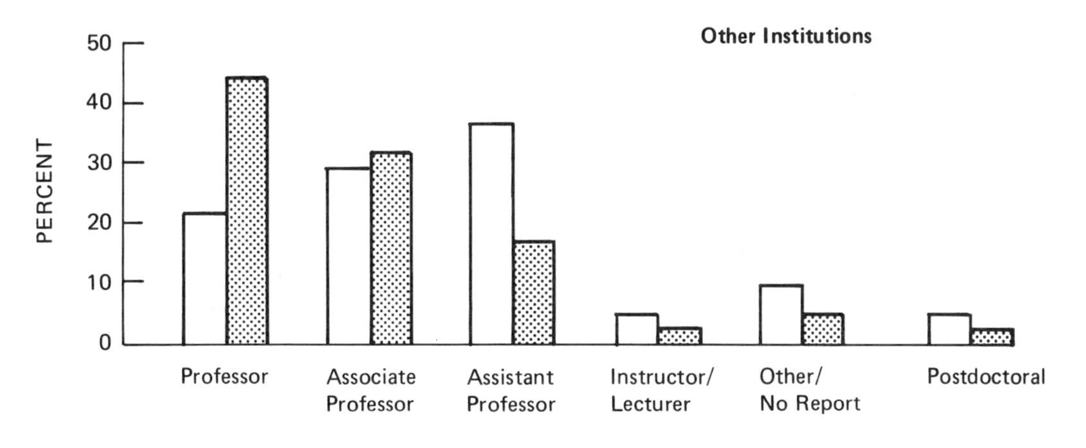

*See Appendix for a description of the ranking of institutions by federal R&D expenditures.

SOURCE: Survey of Doctorate Recipients, National Research Council.

Looking at faculty rank from another perspective we find that in the major research universities, women held 24 percent of the assistant professorships, but only 3 percent of the full professorships in 1981 (Table 4.3). At all ranks, there was some increase since 1977 in the proportion of female faculty. It appears that the substantial increases in women faculty at the assistant professor rank that took place between 1973 and 1977 are now beginning to be evident at the associate professor level.

The 1981 data also show that at the research universities, the percent of women among junior faculty--10 percent in the physical sciences, 26 percent in the life sciences, and 35 percent in the social sciences--matches their share of recent doctorates.

In the physical sciences, the number of women who are full professors at leading institutions is still very low--roughly one or two women per institution for physics, chemistry, and mathematics combined.

Off-ladder positions

Women scientists are still twice or three times as likely as men to hold nonfaculty appointments (Table 4.4). In most fields, the disparity is greater in 1981 than in 1977. In chemistry departments, both the number and the proportion of Ph.D. women in instructor/lecturer positions have increased since 1977, while the figures for men have dropped. Off-ladder appointments are most prevalent for women in chemistry, physics, and mathematics.

This situation appears to hold true in both the major research universities and in other institutions. In the group of colleges and universities which are not among the top 50 in R&D expenditures, women are 13 percent of all Ph.D. employees but as many as 32 percent of those at instructor/lecturer rank (Table 4.5).

Hiring and promotion of junior faculty

In general, women scientists are found in junior faculty positions in proportions exceeding their availability in the recent doctoral pool (Table 4.6). However, these figures do not indicate how many of the female assistant professors are new hires and how many have held that rank for several years. In other words, a relatively high proportion of females currently at assistant professor rank may be symptomatic of either aggressive hiring of recent women Ph.D.s or lack of upward mobility of those hired in the mid 1970s.

This question is illuminated by promotion data which are available for a longitudinal sample of doctorates who responded to the Survey of Doctorate Recipients in both 1977 and 1981, reporting their rank. The resulting statistics show wide sex differences (Figure 4.3). In the

TABLE 4.3 Number and percent of women doctoral scientists and engineers in faculty positions at 50 leading institutions[a] by field, rank, and sex, 1977 and 1981

	Professor		Associate professor		Assistant professor	
	1977	1981	1977	1981	1977	1981
All science and engineering fields						
Total number	14,306	17,253	7,496	7,995	6,455	6,080
Number women	356	543	491	783	1,042	1,428
Women as % of total	2.5%	3.1%	6.6%	9.8%	16.1%	23.5%
Engineering, mathematics, computer sciences, and physical sciences						
Total number	6,677	8,258	3,121	3,337	2,311	2,118
Number women	46	76	64	138	167	212
Women as % of total	1.0%	1.0%	2.0%	4.1%	7.2%	10.0%
Life sciences						
Total number	3,963	4,287	2,274	2,339	1,810	1,912
Number women	169	236	169	276	310	506
Women as % of total	4.3%	5.5%	7.4%	11.8%	17.1%	26.5%
Behavioral and social sciences						
Total number	3,666	4,706	2,101	2,319	2,334	2,050
Number women	141	231	258	369	565	710
Women as % of total	3.8%	4.9%	12.3%	15.9%	24.2%	34.6%

[a]The top 50 institutions by federal R&D expenditures in FY 1980. See Appendix for a listing of the institutions.

SOURCE: Survey of Doctorate Recipients, National Research Council

TABLE 4.4 Number and percent of doctoral scientists and engineers in academe[a] at rank of instructor/lecturer, by field and sex, 1977-1981

	1977					1981				
	Women		Men			Women		Men		
Field	No.	%	No.	%	Δ	No.	%	No.	%	Δ
Mathematics	51	6.1	243	2.4	3.7	49	5.2	143	1.4	3.8
Physics/astronomy	13	4.8	104	1.1	3.7	17	6.3	111	1.2	5.1
Chemistry	59	5.3	290	2.6	2.7	93	7.0	247	2.1	4.9
Biological sci.	93	3.0	180	1.1	1.9	120	2.9	260	1.4	1.5
Psychology	58	2.1	164	1.6	0.5	115	3.0	204	1.8	1.2
Social sciences	67	1.8	190	1.0	0.8	110	2.2	202	1.0	1.2

[a]Included are 2-year and 4-year colleges and universities. Those employed at medical schools are excluded.

SOURCE: Survey of Doctorate Recipients, National Research Council.

TABLE 4.5 Number and percent of women doctoral scientists and engineers in selected positions in academic institutions[a] by R&D expenditures of institution,[b] 1977-1981

	1977		1981	
	Number women	Women as % of total	Number women	Women as % of total
Top 25 institutions				
Total employed in academe	1,992	10.0%	3,005	13.4%
Faculty[c]	1,160	7.5	1,718	9.9
Instructors/lecturers	17	13.6	51	19.4
Postdoctorals	357	18.3	537	27.4
Other/rank not reported	458	19.5	699	24.6
Second 25 institutions				
Total employed in academe	1,074	7.3	1,448	8.9
Faculty	729	5.7	1,036	7.4
Instructors/lecturers	16	23.9	21	28.4
Postdoctorals	159	18.2	151	22.3
Other/rank not reported	170	17.8	240	16.3
Other institutions				
Total employed in academe	9,799	11.0	12,825	12.7
Faculty	8,342	10.2	10,717	11.6
Instructors/lecturers	329	22.4	462	32.4
Postdoctorals	307	22.0	411	24.7
Other/rank not reported	821	22.1	1,235	23.8

[a]Includes 2-year and 4-year colleges and universities. Excludes medical schools.

[b]See Appendix for a description of ranking of institutions by R&D expenditures.

[c]Includes full professors, associate professors, and assistant professors.

SOURCE: Survey of Doctorate Recipients, National Research Council.

TABLE 4.6 Percent women among doctoral scientists and engineers in junior faculty positions[a] by field, 1981

Field	Percent women among assistant professors, 1981	Percent women among supply of recent Ph.D.s[b]
Mathematics	15.2%	13.4%
Physics	6.7	5.6
Chemistry	16.0	13.3
Earth sciences	11.2	9.3
Engineering	3.4	2.6
Agricultural sciences	5.5	7.8
Biological sciences[c]	26.1	25.3
Psychology	41.6	37.9
Economics	9.8	11.4
Sociology/anthropology	40.2	36.9
Other social sciences	23.9	21.4

[a]Includes faculty at 2-year and 4-year colleges and universities. Excludes medical school faculty.

[b]Based on women as a percent of all 1976-1980 Ph.D.s who in 1981 were employed or seeking employment. The 1976-1980 Ph.D.s were selected as the appropriate pool since this cohort represents the majority (percent) of doctoral scientists at the assistant professor rank.

[c]See footnote a.

SOURCE: Survey of Doctorate Recipients, National Research Council.

FIGURE 4.3 Promotions of doctoral scientists and engineers in junior faculty positions between 1977 and 1981

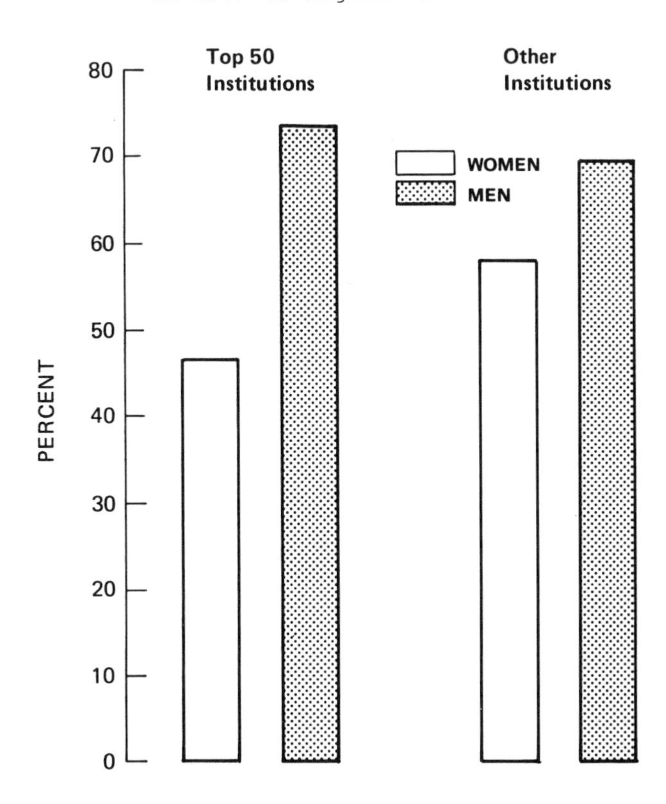

See Appendix C for a description of ranking of institutions by federal R&D expenditures.

SOURCE: Survey of Doctorate Recipients, National Research Council

NOTE: Based on survey responses in 1977 and 1981 from a common set of individuals. The sample sizes for women and men who were assistant professors in 1977 and who reported rank in 1981 are (a) in the top 50 institutions, 125 women and 200 men; and (b) in other institutions, 411 women and 601 men. For both men and women, the median year of receipt of the doctorate was 1973.

group of top 50 institutions (ranked by R&D expenditures), for example, three-fourths of the men, but only one-half of the women were promoted from assistant professor to a higher rank between 1977 and 1981. The differential in promotion rates is not as large for the group of "other" institutions but remains statistically significant.

Sex differences in promotions of junior faculty remained even when males and females were closely matched by years since doctorate, subfield, and the prestige of the department from which they received their Ph.D. Women were also found to lag behind men in faculty promotions regardless of marital status and presence of children and whether their work orientation was primarily research or primarily teaching.[6]

Female assistant professors are also more likely to report that their positions are not tenure-track. The difference is 15 percent for women versus 11 percent for men in the leading universities but grows to 25 percent for women and 14 percent for men in other institutions (Table 4.6A).

Tenure decisions

Overall, female scientists in senior faculty positions are less likely than male faculty members to hold tenured positions. At the associate professor rank, 76 percent of the women have received tenure compared with 83 percent of the men (Table 4.7). The differential has narrowed since 1977, however, when the figures were 71 and 82 percent, respectively.

A more detailed comparison of tenure status for associate professors is presented in Figure 4.4. Physics and biochemistry departments as a group show sizeable differentials in awarding of tenure.

At the assistant professor rank, a slightly higher percent of the women have tenure (Table 4.7). It should be noted, though, that achieving tenure at the assistant professor rank is institution-dependent; it is more common in other (not leading) institutions which employ proportionately more women.

The male-female tenure comparisons shown here for science and engineering are similar to those for the total population of college and university faculty.[7]

---

[6]Ahern and Scott, 1981.

[7]Chronicle of Higher Education, September 30, 1981. Based on 1980-81 data from the National Center for Education Statistics for faculty in all fields (including non-Ph.D.s), the proportion with tenure was: associate professors, 83.2 percent for men and 81.9 percent for women; and assistant professors, 26.6 percent for men and 30.5 percent for women.

TABLE 4.6A Percent of assistant professorships that are off-ladder positions, for male and female doctoral scientists and engineers, 1981

	Number of assistant professors		% Who are not tenure-track[b]	
	Women	Men	Women	Men
All 4-year colleges and universities	5,826	20,882	22.5	13.7
Top 50 institutions[a]	1,428	4,652	15.0	11.0
Other institutions	4,398	16,230	24.5	14.3

[a]See Appendix C for a listing of institutions categorized by federal R&D expenditures.

[b]Based on total reporting tenure status.

SOURCE: Survey of Doctorate Recipients, National Research Council

TABLE 4.7 Tenure status of science and engineering faculty at 4-year colleges and universities[a] by rank and sex, 1977 and 1981

	Number and percent[b] in tenured positions:							
	1977				1981			
	Number		% Tenured		Number		% Tenured	
Faculty rank	Women	Men	Women	Men	Women	Men	Women	Men
Professor	2,314	49,275	92.0	95.8	3,049	57,865	92.5	96.0
Associate professor	2,638	29,784	71.4	81.6	3,860	30,060	75.8	82.6
Assistant professor	593	3,458	10.0	12.6	670	2,062	9.7	8.4

[a] Includes medical schools.

[b] Percent is based on the number who reported tenure status.

SOURCE: Survey of Doctorate Recipients, National Research Council.

If we consider only those faculty who have already received tenure, the median time-to-tenure is somewhat shorter for women than for men, 5.9 years versus 6.1 years overall (Table 4.8). For both sexes time-to-tenure, based on number of years since receipt of the doctorate, is shortest in the behavioral and social sciences and in "other" institutions. In the engineering, math, and physical science fields, the typical awarding of tenure appears to lag for female scientists, one-fourth of whom did not become tenured until 11 or more years after the doctorate.

Administrative positions

The term "administrator" represents a myriad of actual job titles, which for faculty members can cut across all ranks. Academic scientists and engineers who report "administration" as their primary activity may include assistant deans as well as deans, directors of foreign student affairs, affirmative action officers, and persons in other assorted positions. Considering administration generally and controlling for years since doctorate (Table 4.9), we find similar proportions of male and female scientists so employed with one important exception: Relatively few of the senior women in the large group of "other" institutions hold administrative jobs.

Other sources indicate that the total number of women in administrative jobs at colleges and universities has shown modest gains in a recent three-year period, increasing approximately 6 percent per year. As of 1978-79, women administrators were being paid less than men in the same positions, and their lower salaries could not be explained by reasons frequently cited: shorter length of service, having been hired from within the institution, or financial exigencies. And interestingly enough, academic administrative positions with the highest rate of job openings did not appear to be held by increasingly large percentages of women or minority-group members.[8]

[8] "Despite Gains, Women, Minority-Group Members Lag in College Jobs," *Chronicle of Higher Education*, February 3, 1982, and *Women and Minorities in Administration of Higher Education Institutions*, College and University Personnel Association, 1981. The figure of 5.7 percent annual increase was calculated from data reprinted in the *Chronicle* article.

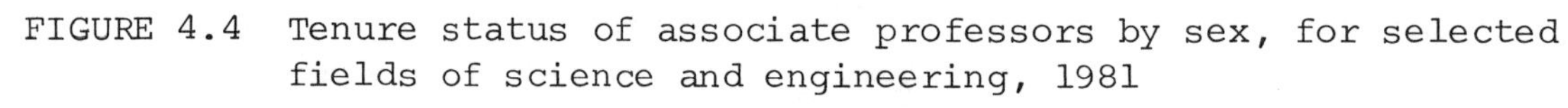

FIGURE 4.4 Tenure status of associate professors by sex, for selected fields of science and engineering, 1981

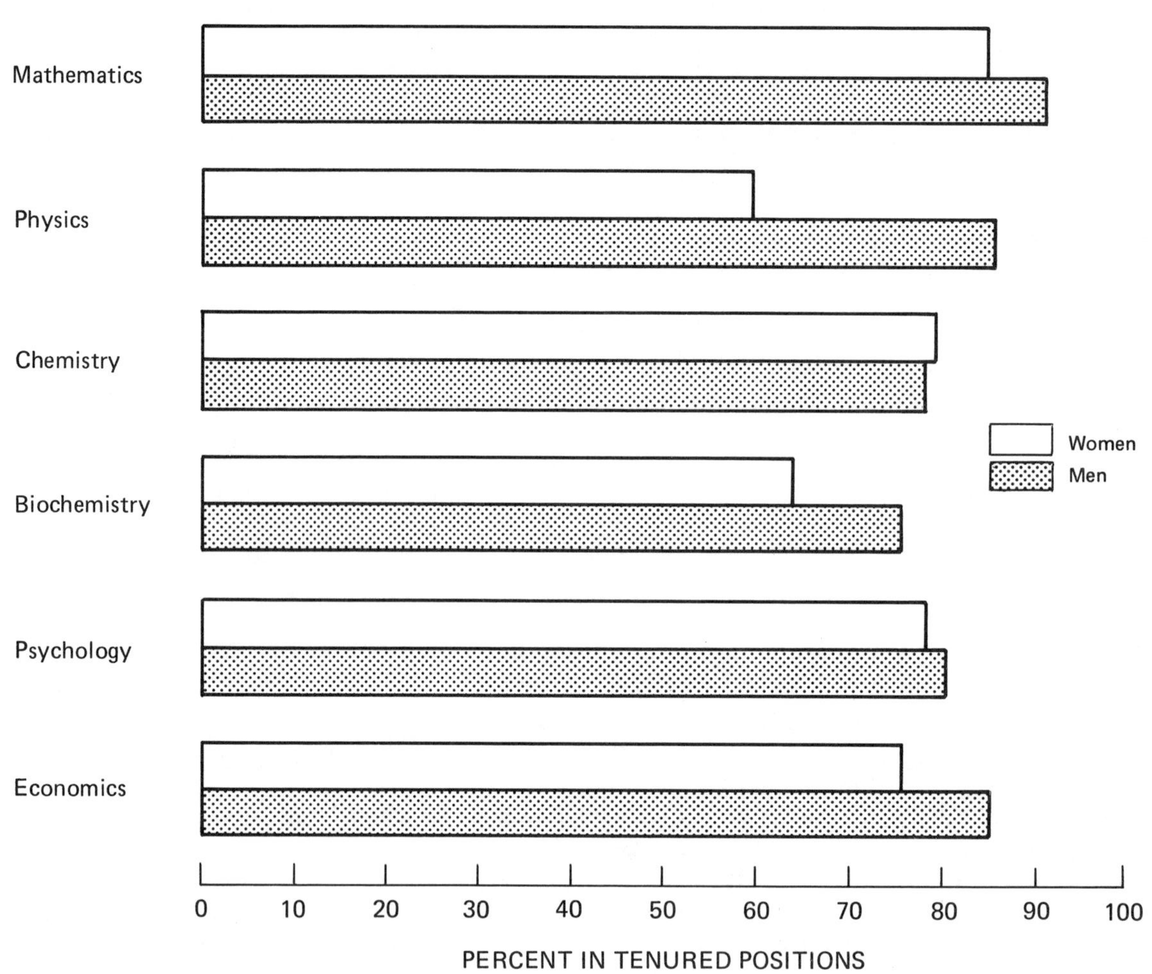

TABLE 4.8 Elapsed time from Ph.D. to tenure for doctoral scientists and engineers in faculty positions by R&D expenditures of institution,[a] field, and sex, 1981[b]

	25th Percentile		50th Percentile		75th Percentile	
	Women	Men	Women	Men	Women	Men
All institutions and fields	3.9	4.1 yrs.	5.9	6.1 yrs.	8.8	8.6 yrs.
Top 25 institutions by R&D	4.3	4.9	6.5	6.4	9.1	9.1
Second 25 institutions by R&D	4.8	4.5	6.6	6.4	10.2	8.9
Other institutions	3.7	3.9	5.7	5.9	8.7	8.4
Engineering, mathematics, computer sciences, and physical sciences	4.9	4.5	6.9	6.3	10.8	8.7
Life sciences	3.8	4.5	6.4	6.3	9.3	8.9
Behavioral and social sciences	3.6	3.5	5.4	5.5	8.0	8.1

[a]See Appendix C for a description of ranking of institutions by federal R&D expenditures.

[b]Includes only those who had been awarded tenure as of 1981. The percent of faculty not yet tenured was 51 percent for women and 26 percent for men.

SOURCE: Survey of Doctorate Recipients, National Research Council.

TABLE 4.9 Number of doctoral scientists and engineers in academic administration, by R&D expenditures of institution,[a] Ph.D. cohort, and sex, 1981

	Pre-1960 Ph.D.s		1960-1969 Ph.D.s		1970-1980 Ph.D.s	
	Women	Men	Women	Men	Women	Men
Top 50 institutions						
Total employed[b]	441	9,704	888	12,235	2,841	12,400
Number administrators	63	1,543	110	1,686	177	846
Percent administrators	14.3	15.9	12.4	13.8	6.2	6.8
Other institutions						
Total employed	1,306	16,917	3,044	33,711	8,556	40,400
Number administrators	85	2,481	399	4,562	575	2,314
Percent administrators	6.5	14.7	13.1	13.5	6.7	5.7

[a]See Appendix C for a description of ranking of institutions by federal R&D expenditures.

[b]Excludes postdoctorals

SOURCE: Survey of Doctorate Recipients, National Research Council

Faculty salaries

Salaries of male and female faculty members have been the subject of numerous studies by individual researchers, professional societies, and education associations. There is wide agreement that women as a group are paid less than men at the same rank. The results differ in the magnitude of the pay differential, depending on how the salaries are disaggregated by field, cohort, type of institution, etc. The first report of this Committee concluded that sex differences in salaries remain a serious problem in academic institutions. At the full professor level, the differences as of 1977 amounted to at least $2,500 between the median salaries paid to men and women, reaching a dollar difference of $6,200 in chemistry.

A subsequent study was carried out to ascertain whether the salary differences (and also rank and tenure differences) diminish or disappear when the male and female faculty members are closely matched. In this case, male-female pairs in a sample were matched by year of Ph.D., field of Ph.D., the granting institution, total full-time equivalent years of professional experience, and race. At each rank and for each cohort, the lower median salaries for women faculty members persisted. In fact, the previously-mentioned salary deficit of $6,200 for female chemistry professors was reduced only to $5,500.[9]

The median salaries shown in Table 4.10 are the most recent available data on a national sample of Ph.D.s. For some field-rank categories, there were too few sample individuals to provide meaningful statistics; the median salaries are not provided in these cases.

After controlling for rank, salary differences for men and women persist in most fields.[9a] At full professor rank, the differentials amount to $1,000 to $6,000, depending on the field. The salary deficits continue to be largest in chemistry and medical sciences, and are of the same order of magnitude as they were in 1977. Economics is the third field with large pay differences although we do not have the earlier data for comparison.

Further examination of the data reveals that in chemistry, the salary gap for full professors may be explained in part by the lack of women at the top departments; this is not true to the same extent in medical sciences.

For associate professors, the sex difference in median salary ranges up to $2,500 annually. At the assistant professor rank, male and female scientists appear to receive comparable salaries, especially in physics, computer sciences, and the social sciences.

---

[9] Ahern and Scott, 1981.

[9a] Rank may itself be subject to bias. The rank distributions of female and male faculty are examined on pp 4.5-4.8.

TABLE 4.10 Median annual salaries[a] of doctoral scientists and engineers in faculty positions at 4-year colleges and universities[b] by field of employment, rank, and sex, 1981

Field	Number in sample reporting salary		Median salary[a]		$ Difference	% Difference
	Women	Men	Women	Men		
FULL PROFESSORS						
Mathematics	100	378	$34,700	$36,700	-$2,000	- 5%
Computer sciences	7	42	---	38,200	---	---
Physics	37	209	37,000*	38,000	- 1,000	- 3
Chemistry	75	204	31,600	36,000	- 4,400	-12
Earth sciences	13	249	---	39,600	---	---
Engineering	12	276	---	41,200	---	---
Agricultural sciences	9	372	---	36,000	---	---
Medical sciences	49	150	35,900*	42,100	- 6,200	-15
Biochemistry	6	58	---	40,700	---	---
Other biosciences	107	602	34,100	36,700	- 2,600	- 7
Psychology	130	281	34,500	37,700	- 3,200	- 9
Economics	38	101	36,400*	42,200	- 5,800	-14
Sociology/anthropology	79	101	34,700	37,300	- 2,600	- 7
Other social sciences	91	239	34,800	36,800	- 2,000	- 5
ASSOCIATE PROFESSORS						
Mathematics	129	218	27,800	28,600	- 800	- 3
Computer sciences	28	82	29,800	30,000	- 200	- 1
Physics	24	117	---	28,600	---	---
Chemistry	83	123	27,000	28,700	- 1,700	- 6
Earth sciences	31	125	28,300	29,400	- 1,100	- 4
Engineering	19	126	---	32,700	---	---
Agricultural sciences	14	163	---	29,500	---	---
Medical sciences	75	95	30,200	32,800	- 2,600	- 8
Biochemistry	8	39	---	29,800	---	---
Other biosciences	139	288	27,000	28,300	- 1,300	- 5
Psychology	136	165	27,100	27,200	- 100	---
Economics	38	44	30,100	32,700*	- 2,600	- 8
Sociology/anthropology	79	70	26,000	27,600	- 1,600	- 6
Other social sciences	102	132	28,100	27,400	- 700	- 3

TABLE 4.10 (cont.) Median annual salaries[a] of doctoral scientists and engineers in faculty positions at 4-year colleges and universities[b] by field of employment, rank, and sex, 1981

Field	Number in sample reporting salary		Median salary[a]		$	%
	Women	Men	Women	Men	Difference	Difference
ASSISTANT PROFESSORS						
Mathematics	130	129	$22,400	$22,600	- $200	- 1%
Computer sciences	31	45	27,900*	27,900		---
Physics	31	68	22,500*	23,300	- 800	- 3
Chemistry	79	90	20,700	22,400	- 1,700	- 8
Earth sciences	40	97	22,600	23,800	- 1,200	- 5
Engineering	29	77	28,100	27,700	+ 400	+ 1
Agricultural sciences	35	105	23,600	24,600	- 1,000	- 4
Medical sciences	50	63	25,100	28,000	- 2,900	-10
Biochemistry	14	18	---	---	---	---
Other biosciences	150	193	22,000	23,200	- 1,200	- 5
Psychology	143	87	22,400	21,600	+ 800	+ 4
Economics	22	41	---	24,100	---	---
Sociology/anthropology	68	57	21,800	22,200*	- 400	- 2
Other social sciences	78	105	22,000	22,400	- 400	- 2

[a] For full-time employees only. Academic salaries for 9-10 months have been adjusted to 12-month equivalents. Salaries are not provided for categories in which the sample size was fewer than 25. Salaries that are asterisked have estimated sampling errors exceeding ± $1,000.

[b] Excluding medical schools

[c] Four fields showed a difference of more than $1,000 in estimated median salaries of male and female assistant professors: chemistry, earth sciences, medical sciences, and "other" biosciences. The results for medical sciences may be explained by sex differences in time-in-service, as approximated by time since receipt of the doctorate. In the other three fields, however, the female assistant professors had, on the average, a longer elapsed time since the doctorate than the male assistant professors.

SOURCE: Survey of Doctorate Recipients, National Research Council.

An attempt was made to examine the salary data in greater detail according to type of institution, but with this additional break-out the numbers of sample cases were too small to permit meaningful statistics.

Geographic mobility

Only a small proportion (5 percent in 1978-79)[10] of all science and engineering faculty members switch employers in a given year. Still, career advancement among academics is believed to be tied to the ability or willingness to relocate oneself and perhaps one's family to accept a more desirable position at another institution. It is also presumed that women faculty are more likely to be constrained in geographic location because of a spouse's career, partially explaining their slower career progress.

There is some evidence that geographic constraints are more frequently acknowledged by married women than by married men. This was evident at the earliest career stages, at the time of deciding whether to take a postdoctoral position (see Chapter 3). One study (Marwell, Rosenfeld, and Spilerman, 1979) which unfortunately was based on 10-year old data, indicated that academic women are less likely than men to change geographic area when they change jobs. Other data from a 1974 survey show male Ph.D.s with a higher "mobility index" based on actual moves over a 10-15 year period (Ferber, 1978).

In a recent sample of junior faculty, however, women were more likely than their male counterparts to switch institution--whether by choice or by necessity. And the female assistant professors who moved did not materially improve their status while the men who moved did.[11]

It is clear that there is little documentation on the value of geographic mobility to one's long-term career attainment, whether women scientists are in fact less mobile, and whether this makes a difference.

Conclusion

The overall nature of the academic career differences between women and men scientists and engineers has not changed significantly over the 1977-1981 period although the balance among the various factors that define these differences is somewhat altered. There has been marked im-

[10]"Fewer Recent Ph.D.s on Science Faculties," Chemical and Engineering News, February 15, 1982, and Young and Senior Science and Engineering Faculty, 1980, National Science Foundation, 1981.

[11]Ahern and Scott, 1981.

provement in initial hiring of women, who are now appropriately represented at the assistant professor level in line with their greatly increased presence in the pool of recent doctorates. Counterbalancing this finding is a marked trend toward increased occupational sex segregation in academic science; the previous over-representation of women in postdoctoral appointments and off-ladder ranks has increased markedly. In addition, far higher proportions of women than of men hold off-ladder assistant professorships; these are the short-term or temporary replacement positions often characterized as "revolving-door appointments." Promotion, tenure, and salary patterns continue to favor men when factors such as length of experience and institutional category are held constant.

To what extent the relative improvement for women assistant professors in terms of both representation and salaries will ultimately be reflected among senior faculty ranks remains to be seen, but at current promotion rates no significant equalization can be expected for a number of years.

The exceptionally high overrepresentation of women in off-ladder positions remains a matter of grave concern to this Committee. One possible interpretation of this finding is that the situation occurs as the result of women's own choices--unwillingness to relocate, or a preference for a less demanding position while also raising children; others, however, are either that institutions are applying different standards in hiring women faculty or that they continue to enforce covert antinepotism rules.

CHAPTER 5

DOCTORAL WOMEN IN INDUSTRY

A previous report of this Committee examined the comparative status of men and women Ph.D.s employed in industry, basing its findings on 1977 data.[1] In the following section, we reanalyze the industrial science and engineering employment patterns as of 1981.

Increase in employment of female Ph.D.s

In just four years, the number of doctoral women in the business and industry sector doubled--increasing from 1,700 to 3,500 between 1977 and 1981. Growth during the previous 4-year period was at nearly the same rate. Despite these increases, women scientists and engineers represent only 5 percent of all Ph.D.-level personnel in industry (Table 5.1). The fields with the largest numbers of total Ph.D. employment--engineering and chemistry--have relatively few women, 1 percent and 4 percent, respectively.

In several fields, the proportion of women among industrial scientists is lower than their presence in the Ph.D. work force. As mentioned earlier, women are 4 percent of the doctoral chemists employed in industry, but 7 percent of those in all sectors. Similarly, in the biosciences, women account for 11 percent of the industrial personnel, but 18 percent of the Ph.D. supply as a whole. The details of work force availability by field are shown in Table 5.1. In comparison to the pool of recent Ph.D.s (Table 4.6), the underrepresentation is even more marked.

Growth of Ph.D. Personnel

Total employment of doctoral scientists and engineers in industry grew by 14,100 over the 1977-1981 period or at an average annual rate of 5.3 percent. Women represented approximately 1,000 new job-holders or 7 percent of the net growth.

---

[1]Committee on the Education and Employment of Women in Science and Engineering, National Research Council, Women Scientists in Industry and Government: How Much Progress in the 1970s, National Academy of Sciences, 1980.

TABLE 5.1 Percent doctoral women employed in industry and percent available, 1981

Field	Total Ph.D.s in industry	No. women industry	% Women industry	% Women in Ph.D. labor force
All fields	75,629	3,496	5%	12%
Engineering, math & physical sciences	61,102	1,932	3	4
Mathematics	1,117	116	10	8
Computer sciences	4,448	331	7	7
Physics	4,189	109	3	3
Chemistry	19,266	829	4	7
Earth sciences	4,042	167	4	5
Engineering	28,040	380	1	1
Life sciences	9,947	802	8	15
Agricultural sciences	2,545	67	3	3
Medical sciences	3,241	267	8	18
Biological sciences	4,161	468	11	18
Behavioral & social sciences	4,580	762	17	21
Psychology	2,362	550	23	27
Social sciences	2,218	212	10	16

SOURCE: Survey of Doctorate Recipients, National Research Council

Job choices for recent women doctorates

The latest cohort of female doctorate recipients are planning industrial employment at roughly the same rates as their male counterparts. (See Figure 2.3 on page 2.10) This is in contrast to the pattern five years earlier as shown in Table 5.2. For both sexes, an increasing proportion of the Ph.D. graduates are taking jobs in industry but this is more true for women so that the male-female difference has narrowed.

TABLE 5.2 Percent of Ph.D. graduates planning industrial employment following receipt of doctorate by sex for selected fields

	1975 Ph.D.s		1980 Ph.D.s	
	Women	Men	Women	Men
Physics	11	14	25	21
Chemistry	20	30	33	41
Engineering	30	40	42	42

SOURCE: Doctorate Records File, National Research Council.

Type of position held

The available data do not indicate the job titles held by female and male scientists, or the levels of their positions. We do, however, have information on primary work activity and salary and these will be used as a measure of the comparative status of men and women doctorates in industry.

Women scientists are only half as likely as men to hold managerial jobs--16 percent versus 29 percent (Figure 5.1 and Table 5.3). This difference is undoubtedly explained in part by the fact that the women are generally younger. We also find that the relative number of managers among Ph.D. science and engineering personnel has decreased sharply since 1977.

Aside from the lower proportion of women in management, for the most part male and female scientists are distributed similarly by work function (Figure 5.1). Women scientists, however, continue to be relatively overrepresented in basic research and to report "other," not-defined work activities at a higher rate than men.

TABLE 5.3 Percent of doctoral scientists and engineers in industry whose primary work activity is management, 1973, 1977, and 1981

	Women	Men
% Managers		
1973	20.0	40.3
1977	18.1	37.2
1981	16.4	28.7

SOURCE: Survey of Doctorate Recipients, National Research Council

FIGURE 5.1 Primary work activities of doctoral scientists and engineers in industry, 1981

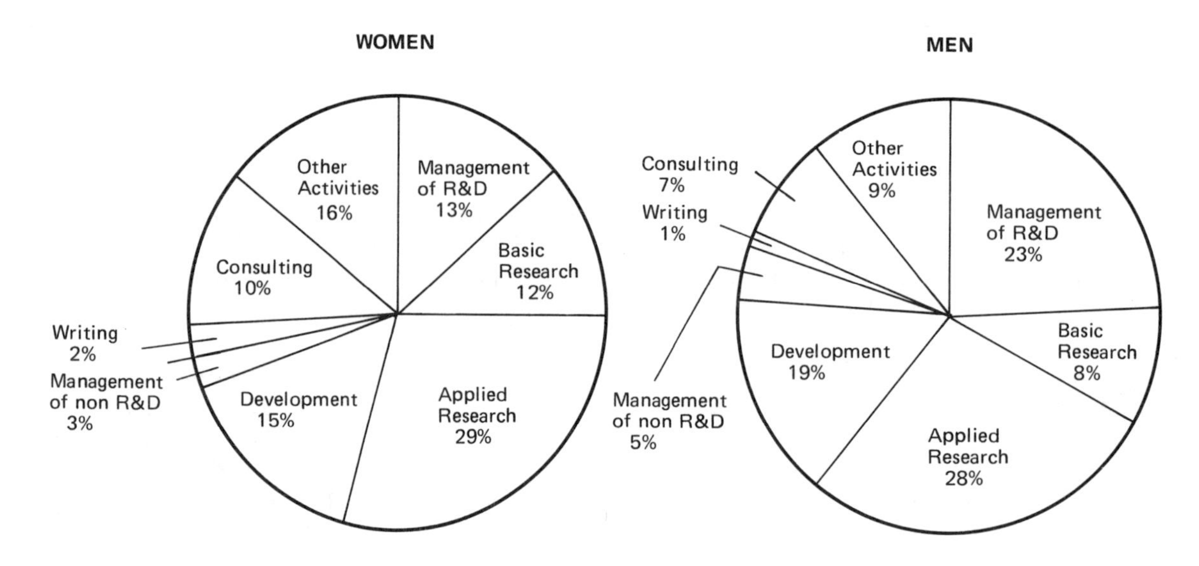

SOURCE: Survey of Doctorate Recipients, National Research Council

Salaries

Median salaries paid to women industrial scientists and engineers are lower than those for men even among the most recent Ph.D.s (Figure 5.2). The sex differential in pay increases with years since doctorate, amounting to $8,000 for those 25 or more years past the doctorate. For the 1979-1980 Ph.D.s, the median female salary was $29,600 or $2,400 less than that for men.

The larger discrepancies for the older doctorates undoubtedly reflect, again, a long past history of discriminatory practice which is in some sense self-perpetuating because these women have not had access to the same opportunities for professional growth. However, it is of concern to find a large pay differential even for the most recent Ph.D.s. We do not know to what extent this may be due to different field distributions for women and men.

FIGURE 5.2 Median salaries of doctoral scientists and engineers in industry by cohort and sex, 1981

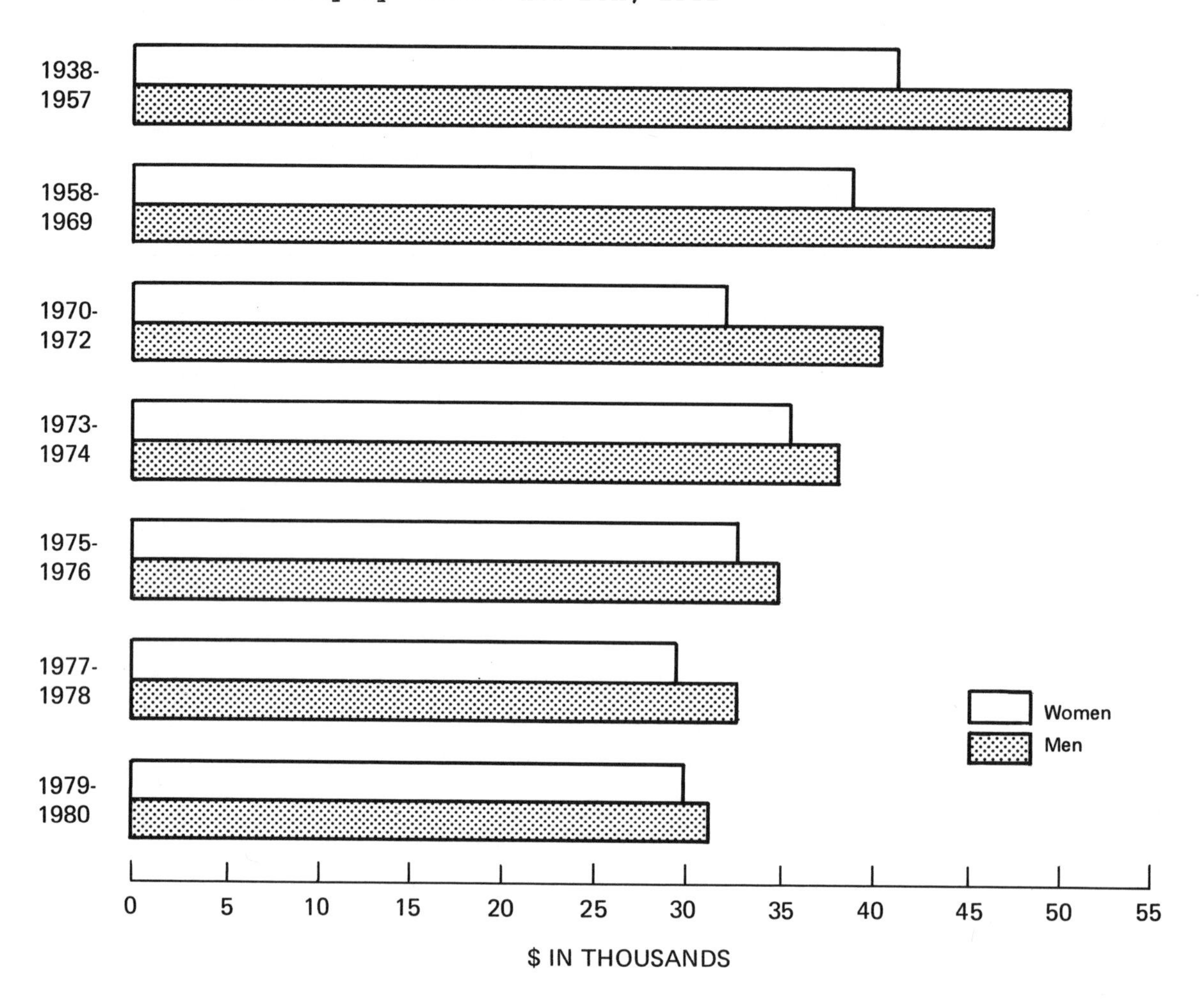

CONCLUSIONS

The most striking change that has taken place since 1977 is the growth in numbers of doctoral women scientists. In just four years, the total supply of women Ph.D.s in science and engineering increased by 50 percent. Ten percent of the present supply is from the 1980 cohort. We also see that among the new doctorates, men and women have very similar characteristics, including the age at which they complete the Ph.D., the type of Ph.D.-granting department, and what their plans are for the immediate future. Males and females in the same field are about equally likely to opt for a postdoctoral appointment, or if they plan to be employed, will go into industry or academe at similar rates.

Once in the academic sector as an employee, however, doctoral women scientists are still far more likely than their male counterparts to be in off-ladder positions. Those who receive appointments as assistant professors appear to have to wait longer to be promoted, on the average. There are, however, signs that the increases in numbers of women scientists among junior faculty that took place between 1973 and 1977 are now being reflected at the associate professor rank. In many fields, sex differences in faculty salaries persist. The salary gaps are largest in the medical sciences, chemistry, and economics, amounting to up to $6,000 for full professors. However, for female assistant professors the salary deficits in several fields have diminished or essentially disappeared since 1977.

Turning to industrial employment, doctoral women scientists and engineers have made numerical gains here as well, more than doubling their numbers from 1,700 to 3,500 in a four-year period. Still, they account for only 5 percent of all Ph.D.-level personnel in industry. They remain less likely to hold managerial jobs, and have lower median salaries than men, even among the most recent Ph.D.s. That recent women graduates are planning industrial employment at the same rate as men is a new phenomenon. By all indications, these younger women scientists believe that their place in the business and industry sector is both wanted and appropriate.

RECOMMENDATIONS

1. That academic institutions, and science departments in particular, give further attention to rates of promotion of male and female faculty members. Much of the focus of affirmative action efforts in the past has been at the entry level. The findings in this report suggest a need to take that one step further and examine the comparative advancement of those hired as assistant professors.

2. That affirmative action policies devote attention to the increasing overrepresentation of women in the off-ladder ranks of instructor/lecturer; although only about 3 percent of women science faculty hold such rank, the trend suggests that institutions have given reduced rather than increased attention to equitable faculty hiring policies.

3. That equal opportunity reviews be linked more directly to departmental or project levels rather than to university-wide performance. Awards should be contingent on satisfactory equal opportunity efforts within the department concerned, rather than requiring evidence of compliance throughout the institution. In restating this recommendation, which was made in the Committee's first report, we are pleased to note that the National Institutes of Health has adopted this policy in their grant reviews, and that the Subcommittee on Women of the Committee on Equal Opportunity in Science and Technology, advisory to the Director of the National Science Foundation, strongly endorses this recommendation.

4. That Congress explore measures to ameliorate the inevitable sex bias in higher education that will again result from a renewed G.I. bill. In raising this question, we emphasize our concern not simply with the numerical disparities that result from a sex-biased support pattern but particularly with the consequent limitations on access for talented women from less than affluent families.

5. That potential new programs aimed at improving the quality of precollege education in mathematics and science be developed with careful regard to the equitable participation of female and minority

students at all levels and from the beginning. In particular, Congressional authorization of such programs and potential state initiatives should include explicit support for such participation in the distribution of public funds.

6. That the National Research Council continue to monitor the progress of women scientists and engineers and to report on this topic at appropriate intervals to the public and to the scientific and engineering professions.

APPENDICES

A. Questionnaire for the 1981 Survey of Doctorate Recipients

B. Questionnaire for the 1980 Survey of Earned Doctorates

C. Top 50 institutions ranked by federal R&D expenditures in fiscal year 1980

D. Sample sizes--Number of doctoral scientists and engineers in academe by field, type of position held, and sex, 1981

APPENDIX A

1981 SURVEY OF DOCTORATE RECIPIENTS

OMB No. 3145-0020

CONDUCTED BY THE NATIONAL RESEARCH COUNCIL WITH THE SUPPORT OF THE NATIONAL SCIENCE FOUNDATION, THE NATIONAL ENDOWMENT FOR THE HUMANITIES, THE NATIONAL INSTITUTES OF HEALTH, AND THE DEPARTMENT OF ENERGY

NOTE: THIS INFORMATION IS SOLICITED UNDER THE AUTHORITY OF THE NATIONAL SCIENCE FOUNDATION ACT OF 1950, AS AMENDED. ALL INFORMATION YOU PROVIDE WILL BE TREATED AS CONFIDENTIAL, WILL BE SAFEGUARDED IN ACCORDANCE WITH THE PROVISIONS OF THE PRIVACY ACT OF 1974, AND WILL BE USED FOR STATISTICAL PURPOSES ONLY. INFORMATION WILL BE RELEASED ONLY IN THE FORM OF STATISTICAL SUMMARIES OR IN A FORM WHICH DOES NOT IDENTIFY INFORMATION ABOUT ANY PARTICULAR PERSON. YOUR RESPONSE IS ENTIRELY VOLUNTARY AND YOUR FAILURE TO PROVIDE SOME OR ALL OF THE REQUESTED INFORMATION WILL IN NO WAY ADVERSELY AFFECT YOU.

If your name and address are incorrect, please enter correct information below.

____

____

____

_ _ _ _ _ - _ _ _ _

INCLUDE NEW NINE-DIGIT ZIP CODE IF KNOWN

(10)

If there is an alternate address through which you can always be reached, please provide it on the line below.

____

c/o Number Street City State ZIP Code (11)

1a. How many full-time equivalent years of professional work experience have you had? _____ **Year(s)** (12-13)

b. Since receiving the doctorate, how many full-time equivalent years of professional work experience have you had? _____ **Year(s)** (14-15)

c. Since receiving the doctorate, how many full-time equivalent years of work experience, if any, involved teaching? _____ **Year(s)** (16-17)

2. What was your employment status (includes postdoctoral appointment*) during FEBRUARY 1981?

☐ Circle your selection and enter number from below (18)

1. Employed full-time (Skip to Question #4)
2. Employed part-time
   If you were employed part-time, were you seeking full-time employment? ☐ Yes ☐ No (19)
3. Postdoctoral appointment*
   If you held a postdoctoral appointment, was it ☐ full-time (Skip to Question #4) ☐ part-time (20)
4. Unemployed and seeking employment
5. Not employed and not seeking employment
6. Retired and not employed

   (4–6: Skip to Question #20)
7. Other, specify ______________

*Temporary appointment in academia, industry or government, the primary purpose of which is to provide for continued education or experience in research.

3. If you were employed part-time during FEBRUARY 1981, what was the MOST important reason for being in that position?

☐ Enter number from below (21)

1. Part-time employment preferred
2. Full-time position not available
3. Constraints due to family or marital status
4. Other, specify ______________

4. From the Degree and Employment Specialties List on page 4 select and enter both the number and title of the employment specialty most closely related to your principal employment or postdoctoral appointment during FEBRUARY 1981. Write in your specialty if it is not on the list.

____

Number Title of Employment Specialty (22-24)

5. If you were employed during FEBRUARY 1981 in a specialty field other than your field of Ph.D., what was the MOST important reason for being in that position?

☐ Enter number from below (25)

1. Better pay
2. More attractive career options
3. Preferred specific geographic location
4. Constraints due to family or marital status
5. Position in Ph.D. field not available
6. Promoted out of position in Ph.D. field
7. Other, specify ______________

6. Please give the name of your principal employer (company, organization, postdoctoral institution, etc. or, if self employed, write "self") and actual place of employment during FEBRUARY 1981.

____
Name of Employer (26-31)

____
Number Street

____
City State

____
ZIP Code (32-40)

7. Which category below best describes the type of organization of your principal employment OR postdoctoral appointment during FEBRUARY 1981?

☐ Enter number from below (41-42)

1. Business or industry (including self-employed)
2. Junior college, 2-year college, technical institute
3. Medical school (including university affiliated hospital or medical center)
4. 4-year college
5. University, other than medical school
6. Elementary or secondary school system
7. Private foundation
8. Hospital or clinic
9. U.S. military service, active duty, or Commissioned Corps, e.g., USPHS, NOAA
10. U.S. government, civilian employee
11. State government
12. Local or other government, specify: ______
13. Nonprofit organization, other than those listed above
14. Other, specify ______

8. What were your primary and secondary work activities during FEBRUARY 1981? (Enter number from the list provided below)

☐ Primary (43-44) ☐ Secondary (45-46)

1. Teaching
2. Basic research
3. Applied research
4. Development of equipment, products, systems, data
5. Design
6. Writing
7. Editing
8. Professional services to individuals

Management or administration of:

9. Research and development
10. Educational programs
11. Other
12. Consulting
13. Production
14. Cultural resources
15. Archival work
16. Curatorial work
17. Performing arts
18. Quality control, inspection, testing
19. Sales, marketing, purchasing, estimating
20. Other, specify ______

9. What was the basic annual salary* associated with your principal professional employment during FEBRUARY 1981? If you were on a postdoctoral appointment (see question #2 for definition), what was your stipend plus allowances? $ ______ per year (47-49)

Check whether salary was for ☐ 9-10 months or ☐ 11-12 months (50)

*Basic salary is your annual salary before deductions for income tax, social security, retirement, etc., but does not include bonuses, overtime, summer teaching, or other payment for professional work.

10a. What was your basic annual salary* for the year ending December 31, 1980? $ ______ per year (51-53)

Check whether salary was for ☐ 9-10 months or ☐ 11-12 months (54)

b. What was your gross professional income† for the year 1980? $ ______ per year (55-57)

†Gross professional income is all payments received for professional activities including basic salary before deductions plus bonuses, consulting fees, honoraria, royalties, rental and subsistence allowances, etc.

11. What percentage of your professional work time did you devote to each of the following activities during FEBRUARY 1981? (Total should equal 100%)

%		
1. ______	(58)	Management or administration of R&D
2. ______	(60)	Management or administration of educational programs
3. ______	(62)	Management or administration of other programs
4. ______	(64)	Teaching
5. ______	(66)	Applied research
6. ______	(68)	Basic research
7. ______	(70)	Consulting
8. ______	(72)	Writing/editing
9. ______	(74)	Development/design
10. ______	(76)	Cultural resources
11. ______	(78)	Other, specify ______

12. If you were employed by an academic institution during FEBRUARY 1981, did you hold a tenured position? 1 ☐ Yes 2 ☐ No (10)

If YES, what year was tenure granted? ______ (11-12)

If NO, did you hold a tenure-track position? 1 ☐ Yes 2 ☐ No (13)

13. If you were employed by an academic institution during FEBRUARY 1981, what was the rank of your position? ☐ Enter number from below (14)

Faculty

1. Professor
2. Associate professor
3. Assistant professor
4. Instructor
5. Administrator
6. Other, specify ______ *Title*

Non-Faculty

7. Teaching staff
8. Research staff
9. Other, specify ______ *Title*

14. Was any of your work during FEBRUARY 1981 supported or sponsored by U.S. Government funds?

1 ☐ Yes 2 ☐ No 3 ☐ Don't Know (15)

If YES, which federal agencies or departments were supporting the work?

Enter number(s) from the List of Federal Supporting Agencies on page 4. ______ (16-39)

15. How important was your DOCTORAL degree in enabling you to attain your present position? (Check only one)

1 ☐ Essential qualification
2 ☐ Helpful, but not essential
3 ☐ Unimportant
4 ☐ Cannot ascertain (40)

16. Listed below are selected topics of national interest. If you devoted a proportion of your professional time which you considered significant to any of these problem areas during FEBRUARY 1981, please give the corresponding number of the ONE on which you spent the MOST time.

☐ Enter number from below (41-42)

1. Energy or fuel
2. Health
3. Defense
4. Environ. protection, pollution control
5. Education (other than teaching)
6. Space
7. Crime prevention and control
8. Food and other agricultural products
9. Natural resources, other than fuel or food
10. Community development and services
11. Housing (planning, design, construction)
12. Transportation, communications
13. Cultural life
14. Other area, specify ______

If you did not select energy or fuel (category #1) in question #16, please skip to question #20.

17. From the list below, give the corresponding number of the ONE energy source that involved the LARGEST proportion of your energy-related work during FEBRUARY 1981.

☐ Enter number from below (43)

1. Coal and coal products
2. Petroleum (including oil shale and tar sands) or natural gas
3. Fission
4. Fusion
5. Hydroenergy
6. Direct solar (including space and water heating, thermal, electric)
7. Indirect solar (winds, tides, biomass, etc.)
8. Geothermal
9. Other, specify ______

18. Please read the following list of energy-related activities and give the corresponding number(s) from the list below of the activity(ies) in which you were engaged during FEBRUARY 1981. Enter number(s) from below ______ (44-63)

1. Exploration
2. Extraction (gas, oil, mining)
3. Manufacture of energy-related components or products
4. Fuel processing (including refining and enriching)
5. Electric power generation
6. Transportation, transmission, distribution of fuel or energy
7. Energy storage
8. Energy utilization, management
9. Fuel reprocessing or disposal
10. Energy conservation
11. Environmental impact (health, economic, etc.)
12. Education, training
13. Research and development
14. Other, specify ______

19. Please enter the number 1-14 from question #18 that BEST describes the activity in which you spent MOST of your energy-related time. ☐ (64-65)

20. What is the major field of your doctorate? Please use the Specialties List on page 4. Please provide the name of the institution where the degree was earned and the year the degree was granted.

______

Ph.D. Field (66-68) Month and Year Granted (69-71) Institution (72-77)

21. Date of Birth

Mo. Day Year

___ ___ ___ (10-14)

22. Citizenship

1 ☐ U.S. Native Born
2 ☐ U.S. Naturalized
3 ☐ Non-U.S., Immigrant (Perm. Res.)
4 ☐ Non-U.S., Immigrant (Temp. Res.) (15)

IF NON-U.S., specify country of citizenship ______ (16-17)

23a. What is your marital status?

1 ☐ Now Married
2 ☐ Widowed
3 ☐ Never Married
4 ☐ Divorced, separated (18)

23b. Do you have any children living with you who are:

Under 6 years of age? 1 ☐ Yes How many? ______ 2 ☐ No (19-20)

Between 6 and 18 years of age? 1 ☐ Yes How many? ______ 2 ☐ No (21-22)

24. Are you physically handicapped? 1 ☐ Yes 2 ☐ No (23) If Yes, enter number(s) from below ______ (24-27)

1. Visual 2. Auditory 3. Ambulatory 4. Other, specify ______

25a. What is your racial background?

1 ☐ American Indian or Alaskan Native
2 ☐ Asian or Pacific Islander
3 ☐ Black
4 ☐ White (28)

25b. Is your ethnic heritage Hispanic?

1 ☐ Yes
2 ☐ No (29)

If Yes, is it:
1 ☐ Mexican-American
2 ☐ Puerto Rican
3 ☐ Other Hispanic (30)

Thank you for completing this questionnaire. Please return the completed form in the enclosed envelope to the Commission on Human Resources, JH638, National Research Council, 2101 Constitution Avenue, Washington, D.C. 20418.

DEGREE AND EMPLOYMENT SPECIALTIES LIST

MATHEMATICAL SCIENCES

000 - Algebra
010 - Analysis & Functional Analysis
020 - Geometry
030 - Logic
040 - Number Theory
052 - Probability
055 - Math. Statistics (see also 544, 670, 725, 727)
060 - Topology
082 - Operations Research (see also 478)
085 - Applied Mathematics
089 - Combinatorics & Finite Mathematics
091 - Physical Mathematics
098 - Mathematics, General
099 - Mathematics, Other*

COMPUTER SCIENCES

071 - Theory
072 - Software Systems
073 - Hardware Systems
074 - Intelligent Systems
079 - Computer Sciences, Other (see also 437, 476)

PHYSICS & ASTRONOMY

101 - Astronomy
102 - Astrophysics
110 - Atomic & Molecular
120 - Electromagnetism
130 - Mechanics
132 - Acoustics
134 - Fluids
135 - Plasma
136 - Optics
138 - Thermal
140 - Elementary Particles
150 - Nuclear Structure
160 - Solid State
198 - Physics, General
199 - Physics, Other*

CHEMISTRY

200 - Analytical
210 - Inorganic
215 - Synthetic Inorganic & Organometallic
220 - Organic
225 - Synthetic Organic & Natural Products
230 - Nuclear
240 - Physical
245 - Quantum
250 - Theoretical
255 - Structural
260 - Agricultural & Food
265 - Thermodynamics & Material Properties
270 - Pharmaceutical
275 - Polymers
280 - Biochemistry (see also 540)
285 - Chemical Dynamics
298 - Chemistry, General
299 - Chemistry, Other*

EARTH, ENVIRONMENTAL AND MARINE SCIENCES

301 - Mineralogy, Petrology
305 - Geochemistry
310 - Stratigraphy, Sedimentation
320 - Paleontology
330 - Structural Geology
341 - Geophysics (Solid Earth)
350 - Geomorph. & Glacial Geology
391 - Applied Geol., Geol. Engr. & Econ. Geol.
395 - Fuel Tech. & Petrol. Engr. (see also 479)
360 - Hydrology & Water Resources
370 - Oceanography
397 - Marine Sciences, Other*
381 - Atmospheric Physics & Chemistry
382 - Atmospheric Dynamics
383 - Atmospheric Sciences, Other*
388 - Environmental Sciences, General (see also 480, 528)
389 - Environmental Sciences, Other*
398 - Earth Sciences, General
399 - Earth Sciences, Other*

ENGINEERING

400 - Aeronautical & Astronautical
410 - Agricultural
415 - Biomedical
420 - Civil
430 - Chemical
435 - Ceramic
437 - Computer
440 - Electrical
445 - Electronics
450 - Industrial & Manufacturing
455 - Nuclear
460 - Engineering Mechanics
465 - Engineering Physics
470 - Mechanical
475 - Metallurgy & Phys. Met. Engr.
476 - Systems Design & Systems Science (see also 072, 073, 074)
478 - Operations Research (see also 082)
479 - Fuel Technology & Petrol. Engr. (see also 395)
480 - Sanitary & Environmental
486 - Mining
497 - Materials Science
498 - Engineering, General
499 - Engineering, Other*

AGRICULTURAL SCIENCES

500 - Agronomy
501 - Agricultural Economics
502 - Animal Husbandry
503 - Food Science and/or Technology (see also 573)
504 - Fish & Wildlife
505 - Forestry
506 - Horticulture
507 - Soils & Soil Science
510 - Animal Science & Animal Nutrition
511 - Phytopathology
518 - Agriculture, General
519 - Agriculture, Other*

MEDICAL SCIENCES

520 - Medicine & Surgery
522 - Public Health & Epidemiology
523 - Veterinary Medicine
524 - Hospital Administration
526 - Nursing
527 - Parasitology
528 - Environmental Health
534 - Pathology
536 - Pharmacology
537 - Pharmacy
538 - Medical Sciences, General
539 - Medical Sciences, Other*

BIOLOGICAL SCIENCES

540 - Biochemistry (see also 280)
542 - Biophysics
543 - Biomathematics
544 - Biometrics and Biostatistics (see also 055, 670, 725, 727)
545 - Anatomy
546 - Cytology
547 - Embryology
548 - Immunology
550 - Botany
560 - Ecology
562 - Hydrobiology
564 - Microbiology & Bacteriology
566 - Physiology, Animal
567 - Physiology, Plant
569 - Zoology
570 - Genetics
571 - Entomology
572 - Molecular Biology
573 - Food Science and/or Technology (see also 503)
574 - Behavior/Ethology
576 - Nutrition & Dietetics
578 - Biological Sciences, General
579 - Biological Sciences, Other*

PSYCHOLOGY

600 - Clinical
610 - Counseling & Guidance
620 - Developmental & Gerontological
630 - Educational
635 - School Psychology
641 - Experimental
642 - Comparative
643 - Physiological
650 - Industrial & Personnel
660 - Personality
670 - Psychometrics (see also 055, 544, 725, 727)
680 - Social
698 - Psychology, General
699 - Psychology, Other*

SOCIAL SCIENCES

700 - Anthropology
703 - Archeology
708 - Communications*
709 - Linguistics
710 - Sociology
720 - Economics (see also 501)
725 - Econometrics (see also 055, 544, 670, 727)
727 - Social Statistics (see also 055, 544, 670, 725)
740 - Geography
745 - Area Studies*
751 - Political Science
752 - Public Administration
755 - International Relations
760 - Criminology & Criminal Justice
770 - Urban & Regional Planning
775 - History & Philosophy of Science
798 - Social Sciences, General
799 - Social Sciences, Other*

HUMANITIES

802 - History & Criticism of Art
804 - History, American
805 - History, European
806 - History, Other*
808 - American Studies
809 - Theater & Theater Criticism
830 - Music
831 - Speech as a Dramatic Art (see also 885)
834 - Philosophy
836 - Comparative Literature
891 - Library & Archival Sciences
878 - Humanities, General
879 - Humanities, Other*

LANGUAGES & LITERATURE

811 - American
812 - English
821 - German
822 - Russian
823 - French
824 - Spanish & Portuguese
826 - Italian
827 - Classical*
829 - Other Languages*

EDUCATION & OTHER PROFESSIONAL FIELDS

801 - Art, Applied
833 - Religion
881 - Theology
882 - Business Administration
883 - Home Economics
884 - Journalism
885 - Speech & Hearing Sciences (see also 831)
886 - Law, Jurisprudence
887 - Social Work
897 - Professional Field, Other*
938 - Education (other than teaching in a field listed above)

899 - Other Fields*

*Identify the specific field in the space on the questionnaire.

LIST OF FEDERAL SUPPORTING AGENCIES (For use with #14)

1. Agency for International Development
2. Environmental Protection Agency
3. National Aeronautics & Space Administration
4. National Endowment for the Arts
5. National Endowment for the Humanities
6. National Science Foundation
7. Nuclear Regulatory Commission
8. Smithsonian Institution
9. Department of Agriculture
10. Department of Commerce
11. Department of Defense
12. Department of Energy
13. National Institutes of Health (DHHS)
14. Alcohol, Drug Abuse & Mental Health Administration (NIAA, NIDA, NIMH)
15. Other DHHS, specify________________
16. National Institute of Education (E.D.)
17. Other Department of Education (E.D.)
18. Department of Housing and Urban Development
19. Department of the Interior
20. Department of Justice
21. Department of Labor
22. Department of State
23. Department of Transportation
24. Other agency or department, specify ________________
25. Don't know source agency

SURVEY OF EARNED DOCTORATES AWARDED IN THE UNITED STATES

APPENDIX B

Conducted by
The National Research Council
in Cooperation with
The American Council of Learned Societies,
The Social Science Research Council, and
The Graduate Deans

Supported by
The National Science Foundation,
The U.S. Office of Education,
The National Endowment for the Humanities, and
The National Institutes of Health

To the Doctoral Candidate:

This is a brief description of the Survey of Earned Doctorates indicating how the resulting data are used and the individual confidentiality of data is protected. The basic purpose of this Survey is to gather objective data about doctoral graduates, data that are often helpful in improving graduate education. We ask your cooperation with the project.

The information requested on the accompanying questionnaire is largely self-explanatory. Please complete it, detach it along the perforated line, and return it to your Graduate Dean. On the back of this sheet is a Specialties List with code numbers and titles for classifying your fields of specialization. This will be useful in connection with several items on the questionnaire. If none of the detailed fields listed seems to be appropriate, note the "General" and "Other" categories.

What is the Survey of Earned Doctorates?

The Survey is conducted annually by the Commission on Human Resources of the National Research Council in cooperation with the American Council of Learned Societies and the Social Science Research Council. The form is distributed with the cooperation of the Graduate Deans and filled out by all graduates who have completed requirements for their doctoral degrees. Research doctorates in all fields are included, but professional degrees such as the MD, DDS, and DVM are not included because information about recipients of those degrees is compiled elsewhere. The cumulative file goes back to 1920 and is called the Doctorate Records File.

The use of the doctoral data has been increasing, partly because of the implications for graduate education stemming from the change in the growth pattern of the number of persons receiving doctorates (562 in 1920; 3,278 in 1940; 9,735 in 1960; 29,497 in 1970; peaking at 33,727 in 1973; and now at 30,850 in 1978). This survey attempts to supply some of the information as of the time the doctorate is received.

What uses are made of the Survey data?

The data collected by this survey questionnaire become part of the Doctorate Records File maintained by the Commission on Human Resources of the National Research Council. The Survey data are collected with the intention that they will be put to use, but only under carefully defined conditions. Such data as the number of degrees awarded in each field of specialization, the educational preparation of degree recipients, their sources of financial support, the length of time required to attain the degree, and postdoctoral employment plans of doctorate recipients are of great interest to graduate schools, employers, the scholarly community, and the nation generally. The Doctorate Records File is used for a limited number of carefully defined follow-up research studies. Each year a sample of doctorate recipients is selected for inclusion in a longitudinal research file maintained for the National Science Foundation, the National Institutes of Health, and the National Endowment for the Humanities.

Statistical summaries from the Doctorate Records File are used by educational institutions, professional societies, and government agencies. Some specific examples are:

- An extensive statistical summary of the data is published and distributed to all graduate schools about every five years.[1] These reports have been widely used by graduate schools and states to evaluate their progress in providing doctoral education. The data may also be useful to graduate students as an aid in selecting a graduate department.
- Annual reports containing statistical summaries based on the most recent year's Survey are distributed to graduate schools, government agencies, and any others on request.[2]

The confidentiality of Survey data is carefully protected.

This information is solicited under the authority of the National Science Foundation Act of 1950, as amended. All information you provide will be treated as confidential and will be used for statistical purposes only. Information will be released only in the form of statistical summaries or in a form which does not identify information about any particular person. There are only two exceptions to this policy: (1) information (name, year, and field of degree) is released to institutions from which you received degrees and to other organizations as part of the address search procedure for follow-up research studies; and (2) information from your form will be made available to the institution where you receive your doctoral degree. Your response is entirely voluntary and your failure to provide some or all of the information will in no way adversely affect you.

(1) National Academy of Sciences, *A Century of Doctorates — Data Analyses of Growth and Change*, Washington, D.C. 1978.
(2) National Academy of Sciences, *Summary Report 1978, Doctorate Recipients from United States Universities*, Washington, D. C. March, 1979.

SPECIALTIES LIST

MATHEMATICS

000 Algebra
010 Analysis & Functional Analysis
020 Geometry
030 Logic
040 Number Theory
050 Probability & Math. Statistics (see also 544, 670, 725, 727, 920)
060 Topology
080 Computing Theory & Practice
082 Operations Research (see also 478)
085 Applied Mathematics
098 Mathematics, General
099 Mathematics, Other*

COMPUTER SCIENCES

079 Computer Sciences* (see also 437)

ASTRONOMY

101 Astonomy
102 Astrophysics

PHYSICS

110 Atomic & Molecular
132 Acoustics
134 Fluids
135 Plasma
136 Optics
138 Thermal
140 Elementary Particles
150 Nuclear Structure
160 Solid State
198 Physics, General
199 Physics, Other*

CHEMISTRY

200 Analytical
210 Inorganic
220 Organic
230 Nuclear
240 Physical
250 Theoretical
270 Pharmaceutical
275 Polymer
298 Chemistry, General
299 Chemistry, Other*

EARTH, ENVIRONMENTAL AND MARINE SCIENCES

301 Mineralogy, Petrology
305 Geochemistry
310 Stratigraphy, Sedimentation
320 Paleontology
330 Structural Geology
341 Geophysics (Solid Earth)
350 Geomorph. & Glacial Geology
391 Applied Geol., Geol. Engr. & Econ. Geol.
360 Hydrology & Water Resources
370 Oceanography
397 Marine Sciences, Other*
381 Atmospheric Physics and Chemistry
382 Atmospheric Dynamics
383 Atmospheric Sciences, Other*
388 Environmental Sciences, General (see also 480, 528)
389 Environmental Sciences, Other*
398 Earth Sciences, General
399 Earth Sciences, Other*

ENGINEERING

400 Aeronautical & Astronautical
410 Agricultural
415 Biomedical
420 Civil
430 Chemical
435 Ceramic
437 Computer
440 Electrical
445 Electronics
450 Industrial
455 Nuclear
460 Engineering Mechanics
465 Engineering Physics
470 Mechanical
475 Metallurgy & Phys. Met. Engr.
476 Systems Design & Systems Science
478 Operations Research (see also 082)
479 Fuel Tech. & Petrol. Engr.
480 Sanitary & Environmental
486 Mining
497 Materials Science
498 Engineering, General
499 Engineering, Other*

AGRICULTURAL SCIENCES

500 Agronomy
501 Agricultural Economics
502 Animal Husbandry
503 Food Science & Technology
504 Fish & Wildlife
505 Forestry
506 Horticulture
507 Soils & Soil Science
510 Animal Science & Animal Nutrition
511 Phytopathology
518 Agriculture, General
519 Agriculture, Other*

MEDICAL SCIENCES

522 Public Health & Epidemiology
523 Veterinary Medicine
526 Nursing
527 Parasitology
528 Environmental Health
534 Pathology
536 Pharmacology
537 Pharmacy
538 Medical Sciences, General
539 Medical Sciences, Other*

BIOLOGICAL SCIENCES

540 Biochemistry
542 Biophysics
544 Biometrics & Biostatistics (see also 050, 670, 725, 727, 920)
545 Anatomy
546 Cytology
547 Embryology
548 Immunology
550 Botany
560 Ecology
564 Microbiology & Bacteriology
566 Physiology, Animal
567 Physiology, Plant
569 Zoology
570 Genetics
571 Entomology
572 Molecular Biology
576 Nutrition and/or Dietetics
578 Biological Sciences, General
579 Biological Sciences, Other*

PSYCHOLOGY

600 Clinical
610 Counseling & Guidance
620 Developmental & Gerontological
630 Educational
635 School Psychology
641 Experimental
642 Comparative
643 Physiological
650 Industrial & Personnel
660 Personality
670 Psychometrics (see also 050, 544, 725, 727, 920)
680 Social
698 Psychology, General
699 Psychology, Other*

SOCIAL SCIENCES

700 Anthropology
708 Communications*
710 Sociology
720 Economics (see also 501)
725 Econometrics (see also 050, 544, 670, 727, 920)
727 Statistics (see also 050, 544, 670, 725, 920)
740 Geography
745 Area Studies*
751 Political Science
752 Public Administration
755 International Relations
760 Criminology & Criminal Justice
770 Urban & Reg. Planning
798 Social Sciences, General
799 Social Sciences, Other*

HUMANITIES

802 History & Criticism of Art
804 History, American
805 History, European
806 History, Other*
807 History & Philosophy of Science
808 American Studies
809 Theatre and Theatre Criticism
830 Music
831 Speech as a Dramatic Art (see also 885)
832 Archeology
833 Religion (see also 881)
834 Philosophy
835 Linguistics
836 Comparative Literature
878 Humanities, General
879 Humanities, Other*

LANGUAGES & LITERATURE

811 American
812 English
821 German
822 Russian
823 French
824 Spanish & Portuguese
826 Italian
827 Classical*
829 Other Languages*

EDUCATION

900 Foundations: Social & Philosoph.
910 Educational Psychology
908 Elementary Educ., General
909 Secondary Educ., General
918 Higher Education
919 Adult Educ. & Extension Educ.
920 Educ. Meas. & Stat.
929 Curriculum & Instruction
930 Educ. Admin. & Superv.
940 Guid., Couns., & Student Pers.
950 Special Education (Gifted, Handicapped, etc.)
960 Audio-Visual Media

TEACHING FIELDS

970 Agriculture Educ.
972 Art Educ.
974 Business Educ.
975 Early Childhood Educ.
976 English Educ.
978 Foreign Languages Educ.
980 Home Economics Educ.
982 Industrial Arts Educ.
984 Mathematics Educ.
986 Music Educ.
987 Nursing Educ.
988 Phys. Ed., Health, & Recreation
989 Reading Education
990 Science Educ.
992 Social Science Educ.
993 Speech Education
994 Vocational Educ.
996 Other Teaching Fields*
998 Education, General
999 Education, Other*

OTHER PROFESSIONAL FIELDS

881 Theology (see also 833)
882 Business Administration
883 Home Economics
884 Journalism
885 Speech & Hearing Sciences (see also 831)
886 Law & Jurisprudence
887 Social Work
891 Library & Archival Science
897 Professional Field, Other*

899 OTHER FIELDS*

* Identify the specific field in the space provided on the questionnaire.

NSF Form 558 1979
OMB No. 99-R0290
Approval Expires June 30, 1981

SURVEY OF EARNED DOCTORATES

This form is to be returned to the GRADUATE DEAN, for forwarding to .... Commission on Human Resources
National Research Council
2101 Constitution Avenue, Washington, D. C. 20418

Please print or type.

1. Name in full: .... (9-30)
(Last Name) (First Name) (Middle Name)

Cross Reference: Maiden name or former name legally changed ....

2. Permanent address through which you could always be reached: (Care of, if applicable) ....

....
(Number) (Street) (City)

....
(State) (Zip Code) (Or Country if not U.S.)

3. U.S. Social Security Number: __ __ __ - __ __ - __ __ __ __ (31-39)

4. Date of birth: .... (Month) (Day) (Year) (10-14) Place of birth: .... (State) (Or Country if not U.S.) (15-16)

5. Sex: 1 ☐ Male 2 ☐ Female (17)

6. Marital status: 1 ☐ Married 2 ☐ Not married (including widowed, divorced) (18)

7. Citizenship: 0 ☐ U.S. native 2 ☐ Non U.S., Immigrant (Permanent Resident)
1 ☐ U.S. naturalized 3 ☐ Non-U.S., Non-Immigrant (Temporary Resident) (19)
If Non-U.S., indicate country of present citizenship .... (20-21)

8. Racial or ethnic group: (Check only one.) *A person having origins in —*
0 ☐ American Indian or Alaskan Native .... any of the original peoples of North America, and who maintain cultural identification through tribal affiliation or community recognition.
1 ☐ Asian or Pacific Islander .... any of the original peoples of the Far East, Southeast Asia,, the Indian Subcontinent, or the Pacific Islands. This area includes, for example, China, India, Japan, Korea, the Philippine Islands, and Samoa.
2 ☐ Black, not of Hispanic Origin .... any of the black racial groups of Africa.
3 ☐ White, not of Hispanic Origin .... any of the original peoples of Europe, North Africa, or the Middle East.
4 ☐ Puerto Rican .... Puerto Rico, regardless of race.
5 ☐ Mexican-American .... Mexico, regardless of race.
6 ☐ Other Hispanic .... Central or South America, Cuba, or other Spanish culture, regardless of race. (22-24)

9. Number of dependents: Do not include yourself. (Dependent = someone receiving at least one half of his or her support from you) .... (25)

EDUCATION

10. High school last attended: .... (26-27)
(School Name) (City) (State)

Year of graduation from high school: .... (28-29)

11. List in the table below all collegiate and graduate institutions you have attended including 2-year colleges. List chronologically, and include your doctoral institution as the last entry.

Institution Name	Location	Years Attended		Major Field		Minor Field	Degree (if any)		
				Use Specialties List			Title of Degree	Granted	
		From	To	Name	Number	Number		Mo.	Yr.

12. Enter below the title of your doctoral dissertation and the most appropriate classification number and field. If a project report or a musical or literary composition (not a dissertation) is a degree requirement, please check box. ☐ (12)

Title ....

Classify using Specialties List

....

Number Name of field

....

13. Name the department (or interdisciplinary committee, center, institute, etc.) and school or college of the university which supervised your doctoral program: ....
(Department/Institute/Committee/Program) (School)

14. Name of your adviser for dissertation, project report or music/literary composition: ....
(Last Name) (First Name) (Middle Initial)

continued on next page

SURVEY OF EARNED DOCTORATES, Cont.

15. Please enter a "1" beside your primary source of support during graduate study. Enter a "2" beside your secondary source of support during graduate study. Check (√) all other sources from which support was received.

a — NSF Fellowship
b — NSF Traineeship
c — NIH Fellowship
d — NIH Traineeship
e — NDEA Fellowship
f — Title IX Graduate & Professional Opportunities Pgm. Fellowship
g — Other HEW
h — AEC/ERDA/DOE Fellowship
i — NASA Traineeship
j — GI Bill
k — Other Federal support (specify) ....
l — Woodrow Wilson Fellowship
m — Other U.S. national fellowship (specify) ....
n — University Fellowship
o — Teaching Assistantship
p — Research Assistantship
q — Educational fund of industrial or business firm
r — Other institutional funds (specify) ....
s — Own earnings
t — Spouse's earnings
u — Family contributions
v — Loans (NDSL direct)
w — Other loans
x — Other (specify) ....

(26-49)

16. Please check the space which most fully describes your status during the year immediately preceding the doctorate.

0 ☐ Held fellowship
1 ☐ Held assistantship
2 ☐ Held own research grant
3 ☐ Not employed
4 ☐ Part-time employed

Full-time Employed in: (Other than 0, 1, 2)
5 ☐ College or university, teaching
6 ☐ College or university, non-teaching
7 ☐ Elem. or sec. school, teaching
8 ☐ Elem. or sec. school, non-teaching
9 ☐ Industry or business
(11) ☐ Other (specify) ....

(12) ☐ Any other (specify) .... (50)

POSTGRADUATION PLANS

17. How well defined are your postgraduation plans?
0 ☐ Am returning to, or continuing in, predoctoral employment
1 ☐ Have signed contract or made definite commitment
2 ☐ Am negotiating with one or more specific organizations
3 ☐ Am seeking appointment but have no specific prospects
4 ☐ Other (specify) .... (51)

18. What are your immediate postgraduation plans?
0 ☐ Postdoctoral fellowship
1 ☐ Postdoctoral research associateship
2 ☐ Traineeship
3 ☐ Other study (specify) ....
} Go to Item "19"
4 ☐ Employment (other than 0, 1, 2, 3)
5 ☐ Military service
6 ☐ Other (specify)......................... (52)
} Go to Item "20"

19. If you plan to be on a postdoctoral fellowship, associateship, traineeship or other study

a. What was the most important reason for taking a postdoctoral appointment? (Check only one.)
0 ☐ To obtain additional research experience in my doctoral field
1 ☐ To work with a particular scientist or research group
2 ☐ To switch into a different field of research
3 ☐ Could not obtain the desired type of employment position
4 ☐ Other reason (specify) .... (53)

b. What will be the field of your postdoctoral study? Please enter number from Specialties List .... (54-56)

What will be the primary source of research support?
0 ☐ U.S. Government
1 ☐ College or university
2 ☐ Private foundation
3 ☐ Nonprofit, other than private foundation
4 ☐ Other (specify) ....
6 ☐ Unknown (57)
Go to Item "21"

20. If you plan to be employed, enter military service, or other—

a. What will be the type of employer?
0 ☐ 4-year college or university other than medical school
1 ☐ Medical school
2 ☐ Jr. or community college
3 ☐ Elem. or sec. school
4 ☐ Foreign government
5 ☐ U.S. Federal government
6 ☐ U.S. state government
7 ☐ U.S. local government
8 ☐ Nonprofit organization
9 ☐ Industry or business
(11) ☐ Self-employed
(12) ☐ Other (specify) .... (58)

b. Indicate what your *primary* work activity will be with "1" in appropriate box; *secondary* work activity (if any) with "2" in appropriate box.
0 ☐ Research and development
1 ☐ Teaching
2 ☐ Administration
3 ☐ Professional services to individuals
5 ☐ Other (specify)........................... (59-60)

c. In what field will you be working? Please enter number from Specialties List .... (61-63)

d. Did you consider taking a postdoctoral appointment? Yes __ No __ (64)

If yes, why did you decide against the postdoctoral?
0 ☐ No postdoctoral appointment available
1 ☐ Felt that I would derive little or no benefit from a postdoctoral appointment
2 ☐ Had more attractive employment opportunity
3 ☐ Other (specify) .... (65)

Go to Item "21"

21. What is the name and address of the organization with which you will be associated?

....
(Name of Organization)

....
(Street) (City, State) (Or Country if not U.S.) (66-71)

BACKGROUND INFORMATION

22. Please indicate, by circling the highest grade attained, the education of

		Elementary school	High school	College	Graduate		
your father:	none	1 2 3 4 5 6 7 8	9 10 11 12	1 2 3 4	MA, MD PhD	Postdoctoral	(72)
your mother	none	1 2 3 4 5 6 7 8	9 10 11 12	1 2 3 4	MA, MD PhD	Postdoctoral	(73)
	0	1 2 3	4 5	6 7	8 9	(11)	

Signature .... Date ....
(74-76)

Top 25 Institutions by Federal R&D Expenditures

1 Johns Hopkins University
2 Massachusetts Institute of Technology
3 University of California, San Diego
4 Stanford University
5 University of Washington
6 University of Wisconsin, Madison
7 Columbia University, Main Division
8 Harvard University
9 University of Michigan
10 Cornell University
11 University of Pennsylvania
12 University of California, Los Angeles
13 University of Minnesota
14 University of California, Berkeley
15 Yale University
16 University of California, San Francisco
17 University of Illinois, Urbana
18 University of Chicago
19 University of Southern California
20 University of Texas at Austin
21 University of Colorado
22 Washington University
23 Pennsylvania State University
24 University of Rochester
25 New York University

Second 25 Institutions by Federal R&D Expenditures

26 Ohio State University
27 California Institute of Technology
28 Purdue University
29 Duke University
30 University of Arizona
31 University of California, Davis
32 Michigan State University
33 University of Iowa
34 Georgia Institute of Technology
35 Northwestern University
36 Texas A & M University
37 University of Utah
38 University of North Carolina at Chapel Hill
39 Yeshiva University
40 Baylor College of Medicine
41 Case Western Reserve University
42 University of Miami
43 Colorado State University
44 University of Connecticut
45 University of Alaska, Fairbanks
46 University of Hawaii, Manoa
47 University of Alabama, Birmingham
48 Woods Hole Oceanographic Institute
49 Oregon State University
50 University of Florida

SOURCE: Survey of Scientific and Engineering Expenditures at Universities and Colleges: Fiscal Year 1980, in Press, National Science Foundation.

Number of doctoral scientists and engineers in academe by field, type of position held, and sex, 1981, showing sample sizes (n) and population estimates (wn)

Field and rank		All colleges & universities[a]			Top 50 institutions by R&D[b]		
		Total	Women	Men	Total	Women	Men
All science and engineering fields							
Total employed	n	11,185	3,749	7,436	3,190	946	2,244
	wn	139,216	17,278	121,938	38,612	4,453	34,159
Total faculty	n	9,625	2,951	6,674	2,446	591	1,855
	wn	123,660	13,471	110,189	31,328	2,754	28,574
Professor	n	4,427	877	3,550	1,275	140	1,135
	wn	59,551	3,232	56,319	17,253	543	16,710
Associate	n	2,932	1,032	1,900	631	197	434
	wn	37,401	4,413	32,988	7,995	783	7,212
Assistant	n	2,266	1,042	1,224	540	254	286
	wn	26,708	5,826	20,882	6,080	1,428	4,652
Instructor	n	204	130	74	31	16	15
	wn	1,762	534	1,228	337	72	265
Other	n	933	473	460	446	215	231
	wn	9,498	2,174	7,324	4,314	939	3,375
Engineering, mathematics, computer sciences, and physical sciences							
Total employed	n	4,668	1,399	3,269	1,307	327	980
	wn	57,660	3,237	54,423	17,107	787	16,320
Total faculty	n	3,939	1,069	2,870	962	185	777
	wn	50,562	2,403	48,159	13,713	426	13,287
Professor	n	1,773	288	1,485	496	31	465
	wn	26,864	647	26,217	8,258	76	8,182
Associate	n	1,213	370	843	241	62	179
	wn	14,901	798	14,103	3,337	138	3,199
Assistant	n	953	411	542	225	92	133
	wn	8,797	958	7,839	2,118	212	1,906
Instructor	n	110	68	42	16	9	7
	wn	741	171	570	102	21	81
Other	n	430	187	243	213	84	129
	wn	4,252	425	3,827	2,047	188	1,859

(cont.) Number of doctoral scientists and engineers in academe by field, type of position held, and sex, 1981, showing sample sizes (n) and population estimates (wn)

Field and rank		All colleges & universities[a]			Top 50 institutions by R&D[b]		
		Total	Women	Men	Total	Women	Men
Life sciences							
Total employed	n	3,596	1,029	2,567	1,253	372	881
	wn	34,654	5,329	29,325	11,239	1,977	9,262
Total faculty	n	3,055	751	2,304	938	213	725
	wn	29,946	3,717	26,229	8,538	1,018	7,520
Professor	n	1,490	209	1,281	498	51	447
	wn	14,084	955	13,129	4,287	236	4,051
Associate	n	888	263	625	250	75	175
	wn	9,030	1,213	7,817	2,339	276	2,063
Assistant	n	677	279	398	190	87	103
	wn	6,832	1,549	5,283	1,912	506	1,406
Instructor	n	46	29	17	9	5	4
	wn	420	140	280	76	22	54
Other	n	278	139	139	166	85	81
	wn	2,404	735	1,669	1,432	457	975
Behavioral and social sciences							
Total employed	n	2,921	1,321	1,600	630	247	383
	wn	46,902	8,712	38,190	10,266	1,689	8,577
Total faculty	n	2,631	1,131	1,500	546	193	353
	wn	43,152	7,351	35,801	9,077	1,310	7,767
Professor	n	1,164	380	784	281	58	223
	wn	18,603	1,630	16,973	4,708	231	4,477
Associate	n	831	399	432	140	60	80
	wn	13,470	2,402	11,068	2,319	369	1,950
Assistant	n	636	352	284	125	75	50
	wn	11,079	3,319	7,760	2,050	710	1,340
Instructor	n	48	33	15	6	2	4
	wn	601	223	378	159	29	130
Other	n	225	147	78	67	46	21
	wn	2,842	1,014	1,828	835	294	541

[a] Excludes medical schools and university-administered national laboratories.

[b] See Appendix C for a listing of the top 50 institutions by federal R&D expenditures in FY 1980.

SOURCE: Survey of Doctorate Recipients, National Research Council

REFERENCES

Ahern, Nancy C., and Elizabeth L. Scott, Career Outcomes in a Matched Sample of Men and Women Ph.D.s: An Analytical Report, National Research Council, National Academy Press, Washington, D.C., 1981.

American Chemical Society, "Women on Chemistry Faculties: Jobs Still Few," reported in Chemical and Engineering News, June 2, 1980, p. 28.

American Physical Society, "Summary of 1980-81 Baranger-Eisenstein Survey Report," reported in Physics Today, February 1982, p. 99.

Armstrong, Jane M., "A National Assessment of Achievement and Participation of Women in Mathematics: A Report to the National Institute of Education," Education Commission of the States, Denver, Colorado, 1979.

Astin, Helen S., Journal of Counseling Psychology, 1968, 15 (6), pp. 536-540.

Bernard, Jessie, Academic Women, 1964, New American Library Edition, p. 48.

Casserly, Patricia Lund, "Helping Able Young Women Take Math and Science Seriously in High School," published by College Board, New York, 1979.

Chemical and Engineering News, "Fewer Recent Ph.D.s on Science Faculties," February 15, 1982.

Chronicle of Higher Education, "Despite Gains, Women, Minority-Group Members Lag in College Jobs," February 3, 1982.

Chronicle of Higher Education, "Fact File, Faculty Tenure Rates 1980-81," September 30, 1981.

Conable, Charlotte Williams, Women at Cornell: The Myth of Equal Education, Cornell University Press, 1977, pp. 19-25.

Earnest, Ernest, Academic Procession, Chapter 6, "High Seriousness in Bloomers," published by Bobbs Merrill, 1953.

Feldman, Saul D., Escape from the Doll's House, Carnegie Commission on Higher Education, McGraw-Hill, New York, 1974.

Fennema, E. and J. Sherman, "Sex-related differences in mathematics achievement, spatial visualization, and socio-cultural factors," in American Education Research Journal, 1977, 14 (1), pp. 51-71.

Ferber, Marianne A. and Betty Kordick, "Sex differentials in the earnings of Ph.D.s," Industrial and Labor Relations Review, 1978, 31 (2), pp. 227-238.

Fox, Lynn H., "The problem of women in mathematics," Report to the Ford Foundation, 1980.

Gould, Stephen Jay, The Mismeasure of Man, New York, 1981, pp. 103-107.

Harmon, Lindsey R., A Century of Doctorates, National Research Council, National Academy of Sciences, Washington, D.C., 1978.

Harris, Ann Southerland, "The second sex in Academe," AAUP Bulletin, 56 (3), pp. 283-295, 1970.

Haven, E. W., "Factors associated with the selection of advanced academic mathematics courses by girls in high school," Research Bulletin, No. 72, p. 12, Educational Testing Service, 1972.

Haven, Elizabeth W. and Dwight H. Horch, How College Students Finance Their Education, College Entrance Examination Board, New York, 1972, p. V.

Helson, Ravena, "Women mathematicians and the creative personality," Journal of Counseling and Clinical Psychology, 1971, 36 (2), pp. 210-2220.

Hier, Daniel B., and William F. Crowley, Jr., New England Journal of Medicine, May 20, 1982, pp. 1202-1205.

Kagan, Jerome, New England Journal of Medicine, May 20, 1982, pp. 1225-1226.

Lockheed, Marlaine, "Sex bias in aptitude and achievement tests used in higher education," in The Undergraduate Woman: Issues in Educational Equity, Pamela J. Perun, editor, Heath, 1982.

Maxfield, Betty D., and Susan M. Henn, Science, Engineering, and Humanities Doctorates in the United States: 1981 Profile, National Research Council, National Academy Press, Washington, D.C., 1982.

McGee, M. G., "Human spatial abilities: Psychometric studies and environmental, genetic, hormonal and neurological influences," Psychological Bulletin, 1979, 86 (5), pp. 889-917.

National Center of Education Statistics, The Condition of Education, 1979.

National Research Council, Committee on the Education and Employment of Women in Science and Engineering, Women Scientists in Industry and Government: How Much Progress in the 1970s?, National Academy of Sciences, Washington, D.C., 1980.

National Research Council, Committee on the Education and Employment of Women in Science and Engineering, Climbing ghe Academic Ladder: Doctoral Women Scientists in Academe, National Academy of Sciences, Washington, D.C., 1979.

National Research Council, Committee on a Study of Postdoctorals and Doctoral Research Staff, Postdoctoral Appointments and Disappointments, National Academy Press, Washington, D.C., 1981.

National Research Council, The Invisible University, Postdoctoral Education in the United States, National Academy of Sciences, Washington, D.C., 1969.

Olson, Keith W., The G.I. Bill, the Veterans, and the Colleges, University Press of Kentucky, Lexington, Ky., 1974, p. 44.

On Campus with Women, Project on the Status and Education of Women, Association of American Colleges, Washington, D.C., No. 31, Summer 1981.

Perun, Pamela J., The Undergraduate Woman: Issues in Educational Equity, Heath, 1982.

Perrucci, Carolyn, "Sex-based Professional Socialization Among Graduate Students in Science," National Research Council, Research Issues in the Employment of Women: Proceedings of a Workshop, National Academy of Sciences, Washington, D.C., 1975, pp. 83-123.

Shields, Stephanie A., "Functionalism, Darwinism, and the Psychology of Women: A Study in Social Myths," American Psychologist, July 1975, pp. 739-754.

Syverson, Peter D., Summary Report 1981: Doctorate Recipients from United States Universities, National Research Council, National Academy Press, Washington, D.C., 1982.